Calculus

The Dynamics of Change

Calculus

The Dynamics of Change

Prepared by the CUPM Subcommittee on
Calculus Reform and the First Two Years

A. Wayne Roberts, Chair
Macalester College

With the special help of four members of the committee
who served as section editors for the four parts of this book

Part I
Sharon Cutler Ross
DeKalb College

Part II
Martin Flashman
Humboldt State University

Part III
Margret Höft
University of Michigan-Dearborn

Part IV
Sheldon Gordon
Suffolk Community College

Published by
The Mathematical Association of America

Library of Congress Catalog Card Number 95-81631

ISBN 0-88385–098-2

Printed in the United States of America

Current printing (last digit):
10 9 8 7 6 5 4 3 2 1

MAA Notes and Reports Series

The MAA Notes and Reports Series, started in 1982, addresses a broad range of topics and themes of interest to all who are involved with undergraduate mathematics. The volumes in this series are readable, informative, and useful, and help the mathematical community keep up with developments of importance to mathematics.

MAA Notes

1. Problem Solving in the Mathematics Curriculum, *Committee on the Teaching of Undergraduate Mathematics*, a subcommittee of the Committee on the Undergraduate Program in Mathematics, *Alan H. Schoenfeld*, Editor
2. Recommendations on the Mathematical Preparation of Teachers, *Committee on the Undergraduate Program in Mathematics, Panel on Teacher Training*.
3. Undergraduate Mathematics Education in the People's Republic of China, *Lynn A. Steen*, Editor.
5. American Perspectives on the Fifth International Congress on Mathematical Education, *Warren Page*, Editor.
6. Toward a Lean and Lively Calculus, *Ronald G. Douglas*, Editor.
8. Calculus for a New Century, *Lynn A. Steen*, Editor.
9. Computers and Mathematics: The Use of Computers in Undergraduate Instruction, *Committee on Computers in Mathematics Education, D. A. Smith, G. J. Porter, L. C. Leinbach, and R. H. Wenger*, Editors.
10. Guidelines for the Continuing Mathematical Education of Teachers, *Committee on the Mathematical Education of Teachers*.
11. Keys to Improved Instruction by Teaching Assistants and Part-Time Instructors, *Committee on Teaching Assistants and Part-Time Instructors, Bettye Anne Case*, Editor.
13. Reshaping College Mathematics, *Committee on the Undergraduate Program in Mathematics, Lynn A. Steen*, Editor.
14. Mathematical Writing, by *Donald E. Knuth, Tracy Larrabee, and Paul M. Roberts*.
15. Discrete Mathematics in the First Two Years, *Anthony Ralston*, Editor.
16. Using Writing to Teach Mathematics, *Andrew Sterrett*, Editor.
17. Priming the Calculus Pump: Innovations and Resources, *Committee on Calculus Reform and the First Two Years*, a subcomittee of the Committee on the Undergraduate Program in Mathematics, *Thomas W. Tucker*, Editor.
18. Models for Undergraduate Research in Mathematics, *Lester Senechal*, Editor.
19. Visualization in Teaching and Learning Mathematics, *Committee on Computers in Mathematics Education, Steve Cunningham and Walter S. Zimmermann*, Editors.
20. The Laboratory Approach to Teaching Calculus, *L. Carl Leinbach et al.*, Editors.
21. Perspectives on Contemporary Statistics, *David C. Hoaglin and David S. Moore*, Editors.
22. Heeding the Call for Change: Suggestions for Curricular Action, *Lynn A. Steen*, Editor.
23. Statistical Abstract of Undergraduate Programs in the Mathematical Sciences and Computer Science in the United States: 1990–91 CBMS Survey, *Donald J. Albers, Don O. Loftsgaarden, Donald C. Rung, and Ann E. Watkins*.
24. Symbolic Computation in Undergraduate Mathematics Education, *Zaven A. Karian*, Editor.
25. The Concept of Function: Aspects of Epistemology and Pedagogy, *Guershon Harel and Ed Dubinsky*, Editors.
26. Statistics for the Twenty-First Century, *Florence and Sheldon Gordon*, Editors.
27. Resources for Calculus Collection, Volume 1: Learning by Discovery: A Lab Manual for Calculus, *Anita E. Solow*, Editor.
28. Resources for Calculus Collection, Volume 2: Calculus Problems for a New Century, *Robert Fraga*, Editor.
29. Resources for Calculus Collection, Volume 3: Applications of Calculus, *Philip Straffin*, Editor.
30. Resources for Calculus Collection, Volume 4: Problems for Student Investigation, *Michael B. Jackson and John R. Ramsay*, Editors.

31. Resources for Calculus Collection, Volume 5: Readings for Calculus, *Underwood Dudley,* Editor.
32. Essays in Humanistic Mathematics, *Alvin White,* Editor.
33. Research Issues in Undergraduate Mathematics Learning: Preliminary Analyses and Results, *James J. Kaput and Ed Dubinsky,* Editors.
34. In Eves' Circles, *Joby Milo Anthony,* Editor.
35. You're the Professor, What Next? Ideas and Resources for Preparing College Teachers, *The Committee on Preparation for College Teaching, Bettye Anne Case,* Editor.
36. Preparing for a New Calculus: Conference Proceedings, *Anita E. Solow,* Editor.
37. Practical Guide to Cooperative Learning, *Nancy Hagelgans, Barbara E. Reynolds, SDS, Keith Schwingendorf, Draga Vidakovic, Ed Dubinsky, Mazen Shahin, G. Joseph Wimbish, Jr.*
38. Models That Work: Case Studies in Effective Undergraduate Mathematics Programs, *Alan C. Tucker,* Editor.
39. Calculus: The Dynamics of Change, *CUPM Subcommittee on Calculus Reform and the First Two Years, A. Wayne Roberts,* Editor.

MAA Reports

1. A Curriculum in Flux: Mathematics at Two-Year Colleges, *Subcommittee on Mathematics Curriculum at Two-Year Colleges,* a joint committee of the MAA and the American Mathematical Association of Two-Year Colleges, *Ronald M. Davis,* Editor.
2. A Source Book for College Mathematics Teaching, *Committee on the Teaching of Undergraduate Mathematics, Alan H. Schoenfeld,* Editor.
3. A Call for Change: Recommendations for the Mathematical Preparation of Teachers of Mathematics, *Committee on the Mathematical Education of Teachers, James R. C. Leitzel,* Editor.
4. Library Recommendations for Undergraduate Mathematics, *CUPM ad hoc Subcommittee, Lynn A. Steen,* Editor.
5. Two-Year College Mathematics Library Recommendations, *CUPM ad hoc Subcommittee, Lynn A. Steen,* Editor.
6. Assessing Calculus Reform Efforts: A Report to the Community, *Alan C. Tucker and James R. C. Leitzel,* Editors.

Preface

Can we, after ten years of calculus reform, say anything definitive about a new course? That is the question that the Calculus Reform and the First Two Years (CRAFTY) Committee began to debate as we looked forward to the 1996 joint mathematics meetings in Orlando.

We thought so, but we did not rush to a conclusion. We were, in our role as a kind of monitor of the movement, aware of most of the projects that had produced materials available to a national audience, and we discussed what major themes were shared. We had visited many sites where new materials and approaches were being implemented, and felt that we could say quite a bit about the problems that must be anticipated if a course is successfully to be changed. But did this add up to a definitive statement of the sort envisioned by the Tulane Conference in 1986 that precipitated all this activity?

We decided that with the help of e-mail, we would try to involve the entire committee in writing an essay to be called *A Modern Course in Calculus*. Such an effort never ends, we discovered, but it does converge after a while. When a reasonable approximation emerged, we began taking it with us to conferences, mailing it to the leaders of major projects, and showing it to colleagues. We did not lack for criticisms, most of them constructive. The resulting essay opens this book. We set it before the larger mathematical community believing that it does capture the basic themes that should characterize a calculus course that is modern in its vision as well as its pedagogy and content.

It was this idea of vision, in fact, that gave rise to the first of the four sections in this book. Sharon Ross, the coordinator of this section, conducted far-ranging interviews with people from across the country who were involved in one way or another with the calculus reform effort, asking them in a variety of ways to describe their vision of what the course should be. Several people who had emerged as national leaders over the past decade were asked to address this same question of vision in full-length essays.

Several visions have been articulated, supporting materials have been developed, and user groups have formed. Still, there are a lot of schools who have held their counsel, some waiting to see if any national consensus develops, some skeptical, some just not sure which of the new approaches was for them. The second section of this book, developed under Martin Flashman's leadership, offers advice to departments contemplating a change.

It is legitimate to wonder if there is any evidence to show that students learn more in the reformed courses than they did in the traditional courses. Part of the trouble in answering such a question stems, of course, from the fact that when goals change, then so do the ways in which we measure achievement. The complexity of these issues are explored in the third section of the book on assessment. Following up on the benchmark set by Lynn Steen when he included samples of final exams in 1987 in *Calculus for a New Century*, Margaret Höft has included in her section a collection of final exams from around the country.

One of the exciting aspects of the reform movement, indeed one of the most powerful arguments for its effectiveness, is the effect it is now having on the rest of the undergraduate mathematics curriculum. In the last section, Sheldon Gordon has explored the effect of calculus reform on precalculus and advanced courses as well as the relationship it has to all of the allied disciplines that use calculus.

It is fitting to say here that the section editors, Sharon Ross, Marty Flashman, Margaret Höft, and Sheldon Gordon, have not only kept to some very arduous deadlines, but have worked as a harmonious team that has made the producing of this book a great pleasure. All of them join me in thanking the leaders of various calculus projects for generously giving their time and energy to our project. The final expression of thanks goes to Arnie Ostebee who, with the assistance of Mary Kay Peterson at St. Olaf College, took on the task of melding the articles and exams, coming off a dizzying array of word processors, into the unified format required for a book.

It has been my privilege to serve on the CRAFTY Committee since its inception just prior to the Tulane Conference, so the publication of this book caps a particularly significant and rewarding decade of my life. It is probably fitting to close with the observation being heard more and more frequently within the committee. Calculus reform, while not a completed task, is certainly a movement now generating its own momentum, and the time has surely come for the committee to shift its attention from CR to AFTY.

Wayne Roberts
Macalester College

Contents

A Modern Course in Calculus

The Context of the Course

The calculus course is at the same time a culmination and a beginning. It is the place where many of the ideas and techniques learned in the secondary mathematics curriculum are pulled together, the place where many of the naturally occurring questions from those courses can be answered in a satisfying way. But it is also the foundation for the study of the natural sciences, engineering, economics, and an ever-increasing number of the social sciences.

The culminating aspect of the calculus course is for many students very real in another sense. It is the last mathematics course they will ever take. The calculus teacher should never forget, therefore, that the course he or she is teaching will fix the impressions of mathematics that many of these future parents, voters, and civic leaders will carry through the rest of their lives.

On the other hand, there are students taking the course who will go on to pursue subjects that make fundamental use of the calculus. Not only must these students have a thorough grounding in calculus, but they need to be encouraged in their interests with some indications of how the subject relates to these interests.

Whatever use students intend to make or actually do make of the calculus course, they should leave the class with some sense of the role that calculus has played in developing a modern world view, the place it holds in intellectual as well as scientific history, and the role it continues to play in scientific work.

Planning the Course

These considerations dictate that the instructor should plan a course as one might plan a self-contained episode of a continuing story. The course should pick up threads from the past, weave them together in a way that brings things to a satisfying conclusion, but shows clearly that there are still exciting episodes to come.

The syllabus of a well-planned course is much more than a list of chapters to be covered; it articulates a theme that is motivated by an engaging introduction, developed with judiciously selected material, and concluded in a way that draws the course into a coherent whole. Above all, its focus is on concepts, helping the student take them apart, understand where they came from, see how their elements are inter-related, and ultimately to see how they might be used in a new context to build insights that are, at least for that individual, new and significant.

The course should get off to a fast start. Students have been told throughout the secondary curriculum that, "You'll see the reason for this when you get to calculus." It's time to make good. The preparation of the students may not be all it could be, but one more review of ideas from elementary mathematics is not as likely to fix deficiencies as it is to lose an audience that was initially excited finally to be in calculus. Get started, and backfill as necessary along the way.

Successful courses frequently begin on day one with an application, a real application, one that students find interesting, relevant to the world of their experience, demanding some real effort to solve. It might be one of the classic problems of calculus: the acceleration of a freely falling body, the swing of a pendulum, the path of a planet (or a satellite). It might be something very contemporary, such as the spread of a disease or a predator/prey model. Those who believe that all calculus problems of significance lead to differential equations start with a problem that ultimately points in that direction. It is not important that the problem be immediately tractable for the students. It is important that it immediately engage their interest, and that they can see that they are indeed moving toward a solution as the course progresses.

There is another person who should find the applications interesting: the instructor. No one can be prepared to discuss two or three applications from every field of interest that might be represented in a class. The instructor should direct attention to applications that he or she is prepared to discuss with both enthusiasm and a mature grasp of the problem being solved. The goal is not so much to solve the problem at hand as it is to understand what features of the problem suggest the methods of calculus, and to learn to look for those same features in new problems.

Many other course structures are also being used successfully. Collaborative learning is the focus for some; extended time projects are the framework for others. Some courses are centered on laboratory experiences, making them similar to a chemistry or biology class. Content changes are the driving force for others; in particular the solving of differential equations is the organizing theme in some modern calculus courses.

The student should feel some ownership of the course. That is, the student should feel that at certain points, he or she was able to take some ideas, work out their implications, and express those implications in his or her own words. The ultimate goal is to instill that confidence that will enable the student to apply the ideas in new, unfamiliar settings.

It is no less true that the teacher should feel some ownership of the course. There is no need to have just one syllabus for all calculus courses, and even in large schools where a contingent of teaching assistants necessarily follow a common outline, ways need to be found to let each of them feel that they are teaching a course that is in some way their own. The best courses bear the personal imprint of a teacher who is excited about introducing students to ideas he or she feels to be at the heart of the material.

The Role of Technology

A calculus course cannot be modernized simply by finding a way to make use of graphing calculators or computers. Neither should a modern course omit these tools where their use contributes to the goals of the course. Spelling checkers on a computer will not make a good writer out of a poor writer, and computer algebra systems will not make a good mathematician out of a poor one, but efficient practitioners of any art will make intelligent use of all the tools that are available.

Significant, real world applications of the type we have called for are frequently tractable precisely because computational help is available. Certain concepts, such as the integral as a limit of sums or the derivative as a slope of a tangent line that is the limit of a sequence of secant lines were once forbidding to illustrate because of the computations involved. These are now easy to illustrate with classroom demonstrations or computing lab assignments and this should be done.

Some concepts, important to the underlying theory that were once very difficult to convey, can, with the aid of technology, be made very real to the students. For example, students have always been slow to accept the idea that a function could be defined as an integral from a fixed point a to the variable point x. Now such a function can be defined, evaluated, graphed, generally explored, and—finally—differentiated. Computers and graphing calculators can be used to literally make graphic many concepts that once had to be grasped at an abstract level, and this too should be done.

On the other hand, technology should not be used to emulate a classic methodology when that same technology might make another method more attractive. For example, a differential equation that happens to be amenable to a certain clever method that can be implemented on a computer algebra system might nevertheless be better solved by numerical methods when a computer is available.

Everyone understands and remembers ideas better when these ideas are discovered as a result of one's personal investigation. The problem with discovery learning is that it takes time. Now, however, technology allows students to perform once forbidding calculations such as finding values of $(1 + h)^{1/h}$ for smaller and smaller values of h, or to observe the effect on the graph of a polynomial of making small changes in the coefficient of one of its terms. Using technology to help students explore, discover, and test conjectures adds greatly to a modern calculus course. The role technology can play in helping students construct knowledge and build conceptual understanding is perhaps more important than its role as a tool of calculation or manipulation.

One question with regard to the use of technology revolves around whether there should be formally structured labs as there are for, say chemistry, or whether labs should simply be places that, like the library, are available as places in which students may find it convenient to work. This question is probably best settled on a campus by campus basis, being related to such things as whether the school serves a primarily commuter or residential student body, whether there is support for lab personnel, whether the lab is accessible and staffed in evening hours, and more.

The case for the use of technology can be summarized this way. In a course where the goal is to teach the calculus, computers or calculators should only be used when there is a sound pedagogical reason for doing so, in which case they certainly should be used. But when used, thought should be given to whether or not they are being used most wisely for the problem at hand. Their proper role is as a tool for experimenting, for discovering, for illustrating, or for substantiating.

That is, they are to be used for developing intuition and insight; they should not be used just so the user can crank out answers to ever larger and more complicated exercises. We agree with the observation that the purpose of computing is insight, not numbers.

Expectations of Students

In an era when international comparisons and pronouncements from U.S. professional societies call for making mathematics accessible to more students, calls for modernizing our courses and using technological aids where appropriate are sometimes heard as calls to be less demanding of students. In fact, the opposite is true. The new courses have higher expectations. They make greater intellectual demands of students, for example, requiring conceptual understanding rather than rote memorization, and rich, genuine problems rather than irrelevant template exercises.

Students of calculus must know certain facts and relationships, even if they are readily available from a calculator or computer algebra system, just as a writer is expected to know facts and relationships readily available from a dictionary or the computer's word processing system. There are circumstances in which failure to utilize a reference or the tools afforded by modern technology would be a mark of incompetence. There are other circumstances, however, in which undue reliance on such tools slows one down or even prevents participation with co-workers who assume a certain level of fluency in one's field.

There really is substantial agreement among mathematicians, scientists, and engineers about what basic facts a student should know about the differentiation and anti-differentiation of certain basic functions or expressions built up from these functions. Questions are sometimes raised about how to be sure that students learn these basic facts in an age when they are encouraged to use calculators on examinations. The analogy with writing seems to provide a sensible answer. When students are given a spelling test, they can't use their dictionaries. When students are quizzed on their differentiation or anti-differentiation facts, they can't use their calculators. Some schools have introduced quizzes of exactly this type, calling them "Gateway Tests." Like spelling tests in a language course, they don't count for a lot in the course grade. It's simply required that a student take them, and keep taking some version of them until mastery of these basic facts is achieved. Such tests send the message that these facts are important, but that they are not the central theme of the course.

Students should move comfortably between symbolic, verbal, numerical, and graphical representation of mathematical ideas. For example, the ability to think of a function in terms of either an algebraic expression or a graphical representation has long been recognized as a valuable skill of mathematical analysis, a skill students may be able to develop more easily with the aid of a computer or graphing calculator. At the same time, students should learn to analyze functional relationships between variables on the basis of observed data or verbal descriptions, and to draw on all four of these modes of representation when thinking about functional relationships or describing them to others.

Drill is necessary to master certain techniques, but it is also important that students encounter problems in a context where more than one technique is needed, and where the techniques needed are not obvious from the placement of the problem in the book. Students must learn how to get started when finding a starting point is part of the problem, and how to sort through a variety of techniques to find one that might work. It is also important to wean students from the answers in the back of the book as the only way to check an answer. Does the answer fall in the range of the initial estimate? Does it pass the test of credibility? Calculus students should develop as part of their thinking process the heuristics of problem solving.

The need for a strategy in attacking a problem is most evident when the problem is large enough, and perhaps ill-defined enough, to admit several starting points, to require the gathering of information not given as part of the problem, to call for a combination of techniques in the solving, and to have an uncertain answer that must be defended as reasonable. Perhaps better described as projects than problems in the context of the course, challenges to estimate the volume of a lake, to describe the growth of a population over time, or to model the course of an epidemic are closer to the kind of problems engineers and scientists must confront. Real problems cannot be measured out in fifty-minute tests and four-minute exercises. Every student of calculus should learn the necessity of living with a problem and eventually having to say something intelligent about its possible solution.

Implicit in the requirement that something intelligent be said is the understanding that it should be said intelligently. Students should be expected in their written work to depend on the clarity, not the charity of the mind of their reader. The clear definitions and standards of crisp reasoning, shorn of all hyperbole, make mathematics an ideal place to learn to write, but beyond that, it is increasingly recognized that writing facilitates learning.

The call for clear definitions and crisp reasoning are properly understood as an indication that groundwork should be laid for the rigor that characterizes mathematical work. While it seems advisable to postpone epsilon-delta proofs, care should be exercised to distinguish between evidence and proof, and to indicate clearly those things that are being left to prove in another course. A student's first understanding of calculus should be intuitive; rigor can come later. Enough has already been said about the role of clear writing so that this statement should not be interpreted as a call to abandon explanations. It is, in fact, the opposite. We call for the student to be able to give a clear explanation of why things work the way they do, what they mean, when they are to be used.

Finally, students should learn in calculus to read technical material, starting with their textbooks. This frequently requires that one use paper and pencil to fill in missing steps or to work out an example. It is a skill that needs to be learned if one is to be able to work in the independent manner expected of a mature student, and calculus is often mentioned as the course in which one gains mathematical maturity.

There are, then, very definite expectations of students, that in reverse order from which we have mentioned them may be summarized as reading, writing that embodies clear reasoning, and problem solving that goes well beyond simply knowing how to calculate. In many applications the actual calculating is quite routine and should be handled easily with the basic facts and techniques students are expected to master. When calculations are more complex, computer algebra systems and calculators are available to help, and in the context of all the expectations of students that have been listed, it can be said that there is little point in stressing complicated calculating skills in calculus.

Expectations of Teachers

Having emphasized what may seem to be the obvious point that there should be clear expectations of students, it seems appropriate to emphasize the equally obvious point that there are expectations of the instructor. We have already said that the instructor must bring to the classroom enthusiasm for mathematics in general, and for calculus in particular. In the absence of such enthusiasm, no amount of planning will create the lively course that calculus ought to be.

On the other hand, enthusiasm without planning is more likely to be thin than lean, and much has already been said about the value of designing a course that has an introduction, a body of material including applications that support a central theme, and a clear culmination. The goal of a good plan is to provide the mathematical depth that leads to genuine understanding, and toward this end some sacrifice might be made in the number of topics covered.

The effective teacher must pay attention not only to what should be taught, but also to how it should be taught. Pedagogical considerations should certainly be informed by research into how students best learn mathematics, and while much remains to be done in this area, there seems to be a developing consensus that contains a strong constructivist message.

Students often learn more effectively when working in groups, and group work can be used in the laboratory, in extended projects, and in at least some—perhaps all homework assignments. A productive learning environment is characterized by a spirit of cooperative achievement rather than competition. Not only has this been established as the way in which many students best learn, but it is certainly a better preparation for the working environment into which most of them will go.

Listening to uninterrupted lectures is for most students not an effective way to learn. Classroom time is better spent posing problems that are interesting and meaningful to students at their current stage of sophistication. The students should then be actively engaged, trying to answer questions and construct the concepts in their own minds. The suggestion that the teacher be not a sage on the stage but a guide on the side has indeed reached the status of cliché in workshops on the teaching of calculus, but students in the classroom will learn more if their minds are actively engaged in trying to answer questions or reformulate concepts in their own terms.

If we agree that students should learn to read actively with pencil and paper at hand and should similarly learn to write clearly, then we must make it part of our responsibility to learn how to teach these skills. We must learn to include student development of these skills in our assessment, and we must plan to respond promptly and constructively to their efforts.

Assessment should, in fact, take into consideration all the expectations of students that have been mentioned. Evaluation must take into account the student's work on extended projects, contributions to a group, appropriate use of technology, technical writing, and independent reading. Progress in these areas may be difficult to assess, but it is not fair to the student to cite these things as expectations, but then to grade only on mastery of computational skills most easily measured on a timed test.

Summary

The calculus course has been, and in all likelihood will continue to be, the central course in the undergraduate mathematics curriculum. From this course the large majority of students will take away their life-long impressions of what mathematics is all about. From this course we will continue to draw future engineers, scientists, and mathematicians. The research done on student learning strongly suggests that if we set high standards and provide proper support, students will fulfill our expectations. The calculus course deserves a substantial investment of resources and effort in any department responsible for undergraduate mathematics instruction.

Part I:

Visions

The opening essay of this volume, *A Modern Course in Calculus*, synthesizes the common threads of the changes in calculus courses and calculus instruction made in the last ten years. How these common threads have been articulated, refined, and tested is the theme of this section. Visions of calculus may sound overblown for an enterprise as prosaic as teaching freshman calculus, but without risk-takers, visionaries, pioneers, as we call them here, reform of calculus never would have happened.

Early on, when calculus projects were first appearing, an element of competitiveness was apparent among project developers. Adrenaline was running and a desire to show my project is better than yours was recognizable. It was as if having lived with an essentially monolithic calculus sequence for so long, we could not envision a world where many and varied calculus sequences were not only possible, but a better paradigm. Now in talking to those pioneers, one is struck by their respect for each others' work and views and their lack of dogmatism. Certainly they do not always agree. In fact at times their opinions are at opposite poles, but their consistent open-mindedness and willingness to learn from others are dominant features.

What follows is the result of interviewing many of the pioneers about early visions of a new calculus sequence and the evolution of those visions. Their thoughts on the main items of *A Modern Course in Calculus* are also included. A list of those participating appears at the end of this piece.

This part of the volume closes with five short essays elaborating on some of the visions of calculus described by the pioneers. In the first, Tom Tucker reflects on the use of multiple representations of concepts. Next, Mai Gehrke and David Pengelley describe the changes in courses and instruction brought about by assigning student projects. In the third essay, John Kenelly discusses the past, present, and future uses of technology in calculus courses. Next, Deborah Hughes Hallett gives advice on constructing a new calculus course that meets the needs of students and faculty in mathematics and in the client disciplines. The concluding essay, by David Smith, examines the issues of changing instructional methods and learning about how our students learn. We hope these essays will be useful to individuals and departments, in particular as part of the planning process outlined in Part III.

Visions of Calculus

Sharon Cutler Ross
DEKALB COLLEGE

Original Visions

The calculus reform pioneers interviewed for this paper describe common dissatisfactions with calculus instruction as their motivation. However, they envisioned different responses. The two most common responses were attempts to enhance the development of conceptual understanding, and to engage students as active participants. For many pioneers, technology, graphing calculators, computer algebra systems, or other tools, offered ways to accomplish these two goals. Not all modern calculus courses incorporate the use of technology as a major component, and technology was never seen as a panacea. Of far greater importance was what was being taught and what was being learned. "The course was deadly." "No learning took place; students passed through the course." "The course needed to be more interesting, more human." "We needed to put meaning back in." One pioneer noted that even though his college's program was generally recognized as successful, when the faculty looked closely at what students knew, it was clear that their success was not substantial. The words memorization, mimicry, templates, meaningless, artificial were used by virtually all the interviewees in describing what had to be changed. Multiple representation of concepts was another early (and current) tool. As one pioneer said, "My sense is that students could do well things that were not valued."

Presenting students with questions that are interesting to them through classroom, laboratory, or project activities is important in all modern calculus courses. Identifying genuine applications generally required discussions with client disciplines, and these in turn led to content changes in the calculus courses. Topics that did not contribute to the understanding of the central themes of calculus or were not critical for the client disciplines were omitted or de-emphasized. Reformed courses are less formal, include more real applications, and focus on the main ideas of calculus. Such courses are sometimes suspected of being watered down, but "understanding shouldn't be confused with rigor" and the absence of algorithms and procedures does not make the course shallow. "Concepts are more important than procedures."

Although at first for most pioneers, the issues of how students learn and how to teach were secondary, others first thought in terms of changing the how rather than the what. The kinds of activities in which students engage were changed to require active involvement by students and their discovery of basic facts. A comparison provided by one pioneer is between freshman calculus and the first computer science course. In CS I, students who may begin with no computer experience finish the term able to create non-trivial working programs. In Calc I, students often added nothing to what they already knew and at best could mimic solutions to template problems.

Early projects typically focused on one idea or approach. For some this meant constructing a new course whose content was centered around the big ideas of rates of change and accumulations and recognized the actual uses of calculus by the client disciplines. As one pioneer put it, "The courses had degenerated into a 'ranch-house' style with all ideas of equal importance and with lots of special cases. We wanted "high-rises" with important concepts identified as being of more importance." These two objectives often resulted in a new emphasis on differential equations and on working with graphs. Although working with graphs eliminated or greatly minimized symbolic representation, this approach was quite different from that of other pioneers whose focus was multiple representations of ideas and concepts. Nonetheless, both approaches share the belief that the old algebra-oriented course cut off many students from understanding both calculus and the ways it is used. The multiple representation focus challenges students to build connections among symbolic, geometric, and numerical descriptions of a situation, and in the process create their own knowledge and understanding. The early, heavy emphasis on graphical representations was a reaction to the almost exclusively algebraic approach. "Graphical and tabular motivations appeared [in texts], but only algebraic approaches were used in the exercises."

Other projects focused on "what calculus has to say to undergraduates." The goals were "to connect mathematics with students," "to get students thinking about mathematics," "to hook students by finding something they're interested in." Student projects and technology were two vehicles used. Projects, that is open-ended, extended-time assignments, are often, but not exclusively, real applications of calculus. Mathematical questions in and of themselves can be of interest to students, and have been successfully employed in modern calculus courses. Technology entered into these courses in several ways. One was to deal with messy real problems and to use calculus as it would be done outside the classroom. Others saw technology's greatest benefit in enabling students to deal with calculus ideas as concrete objects, or to approach mathematics as an empirical science.

Evolution

Original visions about a modern calculus course have evolved as changes were implemented. By far the biggest change in these views has been in the importance of instructional changes. Even pioneers whose first vision included pedagogical changes such as assigning projects, requiring writing about mathematics, and incorporating laboratory experiences found they had under-estimated the extent of the pedagogical changes that have taken place. The one respondent who explicitly had pedagogy as a focus originally soon began using cooperative learning, unthought of at first but found to be "a powerful strategy." Another who was forced by limited resources to create student teams also recognized unexpected benefits of collaborative work. In addition, although problems were originally designed as a way to engage students' interest, this same pioneer discovered constructivism empirically, and this discovery has changed and shaped what instructors do.

Others interviewed also related ways in which initial changes led to instructional changes. For example, adding laboratory experiences or projects to a course based on a traditional text produced uncomfortable situations. The result in many cases was the production of new texts. Students' accomplishments and interests played a major role in shaping these materials. The desire to achieve an early vision of calculus led most pioneers to analyze how they could best help students learn. Some asked for more writing by students, some for less. Other balances were adjusted such as that among graphical, algebraic, and numerical approaches, and that between computational skill building and concept development. If you make real contact with your students and find out all those wonderful things you do are not working, and that students really do not understand, you can run off and hide your head in the sand or announce to the world that the students are not prepared, do not work hard enough, and are generally to blame for the whole mess. Or you can try to find out what is really going when students fail to learn.

Projects that initially planned no content changes often did gradually drop topics, change the emphasis for others, and re-assess topics in other ways. As clutter is cleared away, courses display more integration of topics, big ideas and threads are more easily identified, and overall there is much greater coherence. Although courses had grown presumably in response to demands by the client disciplines, investigation shows the connections between calculus courses and other subjects had nearly all been severed.

Overall, the evolution of original visions has tended to converge. As we gain a better understanding of what helps students learn and use calculus, the relationship among the early visions have become clearer and good ideas are taken up and used by everyone. One pioneer described the evolution by using a triangle with curriculum, technology, and pedagogy, the three major early foci, as the angle labels. Not only is there a vigorous two-way flow along each of the edges, the whole triangle is being compressed to a center where all three areas will be thoroughly integrated.

Surprises

All adventures involve surprises. Nearly all of the surprises encountered in changing the calculus curriculum have been pleasant. Finding out what students could and could not do was a revelation to many. "The students are amazing in the way they take ownership of the mathematics and responsibility for learning." "I was surprised at how good students were at solving traditional problems without understanding, how clever they were at this." "I thought if students could do problems of type x, they understood concept y." "It was a shock that the mathematics in a student's background wasn't available to use in context." One pioneer reported concern that students' wide-spread inability to deal with word problems would doom attempts to deal with real applications. In practice when freed from dependence on algebraic methods, students were able to relax, think, and produce solutions.

Listening to students led naturally to even more changes in instruction. A surprise for some was the extent to which pedagogical issues have become central to a modern calculus course. Pioneers, in general, had not expected that materials and their development would also play such a large role in bringing about change. The development of technology has also impacted instructional change. The speed with which new technology arrived to support new ideas for calculus, especially making numerical and graphical approaches easier than anticipated, was another pleasant surprise. A different aspect of technology use is its effect as an agent of socialization. "The social nature of the classroom changes as does the interaction between students, and this was unexpected."

Another unanticipated benefit was the range of students for whom the new courses were not only accessible but also interesting and useful. Finding genuine applications of mathematics that were interesting and doable by students sometimes led instructors to learn new mathematics as well. A by-product at one school was the resolution of problems presented by differences in student preparation. Not all students embrace the new courses. Students who have done well under the old rules of memorize, mimic, and move on are sometimes not happy to have the rules changed. This area of resistance seems to be shrinking as more high school mathematics courses change. A far larger group appreciate the opportunities modern calculus courses provide for genuine problem solving and deep understanding of calculus concepts.

The extent to which "changes have come as fast as they have," "have exceeded [my] modest expectation," and "how much change has taken place already" was a surprise to many interviewees. The support by numerous cheerleaders for reform and the time commitments made by reformers were also surprising to some. There were unpleasant surprises as well in considering the calculus reform movement. Some of those interviewed remain shocked at the viciousness and personal attacks of reactionaries. They tell of colleagues who refuse to discuss the possibility of change, although some of these same people agree that the current course is bad. The biggest surprise for one pioneer was returning to freshman calculus after not teaching it for several years, and being struck by how bad a course it was. The next surprise was the resistance to doing anything about it. However, some pioneers are encouraged by how easily most new faculty take to teaching a modern calculus course.

But, whether good or bad, "each surprise is a lesson," and the pioneers have tried to learn these lessons.

Lean and Lively?

A question about whether calculus is becoming lean and lively seemed obligatory. Reactions ranged from amusement to mild annoyance. The phrase never captured the depth and breadth of the aims of the calculus reform movement. Surprisingly fewer than half of those interviewed think modern calculus courses are lively, although some said very strongly that they are. These courses are seen as more interesting, though, to both students and instructors. One pioneer is concerned that resistance by students who have been successful in plug-and-chug courses will damp out the liveliness of modern courses.

Although calculus texts are still fat, some techniques and procedures have been dropped from courses. And those that remain often play a very different role than before. "We don't feel obligated to tell [students] everything about calculus." "There have been few content changes, but the syllabus is freeer." As several pioneers noted, calculus is "a large, deep, rich subject; it can't be lean." "There was a misconception that there was a lot of [excess] baggage in calculus." By and large, calculus is still seen as a topics course, but modern calculus courses are organized around the central themes of calculus, resulting in courses that may not be lean but are "less flabby" than before. Refocusing on the principle concepts of calculus brings out the inter-relations among them and tightens the structure of the courses. "Still to come is a common understanding of the goals of calculus." If weight is measured not by textbook mass, but by intellectual demands, new courses may be heavier not leaner. But by engaging students in the study of useful, interesting material with tools that support exploration and problem solving, even courses that require students to think are more successful than procedure-dominated courses. "We've taught sophisticated techniques for students to use on simple problems. We need to reverse this and teach students to use fewer simple techniques on richer problems."

Visions of the Future

It may be that "the jury is still out" on the lean-and-lively issue, but tremendous changes have taken place in the last ten years in calculus instruction. Where do we go from here? Calculus courses are likely to include more modeling, especially using differential equations early on. The situations studied will show more variety as the borders fade between calculus for business majors, life science majors, engineers, *et al.* Classic problems will be re-evaluated; many "have real value and purpose" or did before they each became a section in the textbook. There is a danger of "changing one set of canned applications for another." Closer ties with client disciplines are one way to avoid this. Students who can apply calculus tools may even encourage the use of more mathematics in other fields. The changes in calculus courses will also continue to influence how other mathematics courses will be taught. Reform is well under-way in differential equations, linear algebra, and courses that precede calculus. The Advanced Placement calculus syllabus is under revision. As a reflection of what happens in college courses, the new AP calculus syllabus will be a measure of the institutionalization of modern calculus courses.

Developments in technology will remain a force for change in calculus instruction. It will be increasingly difficult to draw the line between a calculator and a personal computer. Technology will fully support the movement between symbolic, numerical, and graphical representations by making all three links two-way streets. Visual arguments to support conclusions will become more common. Software that makes it possible for students to explore, compute, organize, and report all in one environment will enable students to handle rich, valuable problems more easily.

Although a few pioneers see this as a "critical period" with the movement on "a knife edge," most are optimistic or cautiously so about the future of calculus instruction. "The pendulum will swing some" and "we may likely settle short of the most imaginative proposals." "The *status quo* of ten years ago won't return," and the situation will "never again be as static." "The spread [of calculus reform] is large enough that it may be inevitable." Resistance to substantial change by faculty and students exists in many places, and textbook publishers continue to push for greatest-common-denominator texts that presumably can be sold in more schools. One part of the future of calculus instruction is dealing with the pressures to reverse direction. Too often any problems with a modern calculus course are attributed to the course, while problems in a traditional course are blamed on the students. Reactionaries are often comparing a modern calculus course "with a golden age that never existed." Another component is also already visible, synthesizing what has been learned so far and preparing for second-generation courses.

"We've been moderately successful at getting the discussion [about calculus] started." "More discussion is needed to move forward." The issue most often mentioned for the next round of discussion and development is "helping students learn how to learn on their own." We need to "develop the ability to learn new math so we need not do all topics." "The expansion of information into mathematics will grow;" students must be prepared to deal with this. Another issue is how to integrate understanding and rigor. By focusing on conceptual understanding, we are doing a much better job of showing students why we do mathematics, but the how that distinguishes mathematics from other disciplines; e.g., proof, precise definitions, has not been done as well. Assessment has not received the same attention as other aspects of teaching calculus. Appropriate changes in assessment have been retarded in many places by attempts to compare traditional and modern courses. Assessments used in courses need to match course goals, and most likely, other measures for comparing courses need to be used. Although one interviewee is "proud of the discipline, that we've stayed the course this long," another points out "the enormous challenge to get faculty to think about the goals of the course and [to get] students to learn effectively." "I'm concerned for the mathematics community as a whole that it must reflect what students and clients need." Perhaps as one said, "the healthiest [future] is constant reform, always experimenting so we have a constant Hawthorne effect." "Other more interesting projects are coming. What can we do better?"

Major Themes

The distinguishing features of modern calculus courses are by no means uniform. Some of those most closely identified with changes in calculus instruction are not universally accepted by all pioneers as important, much less as essential. There *is* agreement that students, how and what they learn, are the center of the action. Beyond this agreement the best description may be that offered by one pioneer that probably no one is or should be doing "all these things, but some subset needs to be used." Pioneers repeatedly said no one should be told how to teach; we each need to find own way. What follows are their comments on nine themes most closely identified with calculus changes. Those interested in improving calculus instruction, but who abhor technology, or shudder at the thought of group work, or are put off by some other supposed tenet of calculus reform, can see how pioneers rate each "tenet," and how a modern calculus course that suits them and their students can be constructed. Of course, those who support the *status quo* will be able to find someone who says x is not wonderful for each x. On the other side, "lack of unanimity is not an excuse to not do something."

Multiple representations of concepts is the *sine qua non* for some calculus courses, but certainly not for all. Answers to "Is it essential to incorporate multiple representations?" ranged from a resounding "absolutely" to "important" to "trendy, overstressed" to "how much of a change is this?" "I didn't know there was any other way to do it" was balanced by, "It's the biggest new idea." One pioneer sees it as crucial, because this is "the way students think anyway; it's natural for them." Also it gives a way to "get [calculus] into their minds without words." Another sees multiple representations as a way to bring rigor into the course by asking for heuristic arguments to support conclusions. These are not proofs as would appear in most texts, but they can be used to convey deep understanding of why a conclusion is true. Several pioneers spoke of the importance of the connections between representations. "The nature of the connections is more important." The students "must learn to go from one to another." "The ability to move between representations is critical to mathematical problem solving." Which representations to use is still a topic for discussion. Some would add verbal and tactile to the generally cited numerical, graphical, and symbolic representations. The view that "symbolic is what gives real, deep insight" is disputed by one pioneer who sees "symbolic [representation] as a stumbling block. The three should be numerical, graphical, and verbal" at least in freshman calculus.

The **use of technology** is even less of a defining characteristic. Several pioneers are concerned with the prevalence of the erroneous beliefs that asking students to have a graphing calculator makes a calculus course modern, and that technology is required for a modern calculus course. For most of those interviewed, technology is a "pragmatically essential" tool, one of many students use for exploration and problem solving. "Why do without it?" "It's inevitable, not a choice." Many courses still use technology primarily "to reproduce handwork," but some find technology use of "real benefit in developing concepts" as students "do mathematics in partnership with [this] tool." The graphing capabilities that technology supplies are a major support for using multiple representations of concepts. There is general agreement that technology should be a servant that is not allowed "to drive how we teach," and that is useful for the kinds of activities students need to be doing.

Group work, collaborative learning, or cooperative work is another oft-cited hallmark of a modern calculus course. Having students work in small groups, whether in or out of class, is an enormous change in teaching style for nearly all college mathematics faculty. Is it a necessary change? "Yes, it's an absolute truth that students learn from each other." "The jury is out; there has been some success with it." "It's a worthwhile, valuable, important element." "Working in teams is the real world." "Not clear where it's going; my experiences have been mixed." "It's very close to a necessity; it helps the vast majority of students." "No." In many cases, students working together has emerged naturally from the kinds of assignments they do or from limited availability of computers. Having seen the results of these collaborations, instructors want to produce them in more systematic ways. This is an area where nearly everyone feels the need for training. How comprehensive the use of group work should be remains an open question.

Student projects are often the first place group assignments are made. While they play no role or only a small one in some courses, projects are a key element of other modern calculus courses. Advocates of student projects say that projects are a vital way for students to take ownership of the mathematics, to "tackle real-life, messy problems," to "become really involved." Whether large- or small-scale projects achieve these goals better is not agreed upon. Pedagogical issues such as "how to push students to talk [and] discuss with others," and how to develop good projects are holding back some who see project experience as valuable and interesting for students. As more instructors work out the logistics of using projects, and more projects become available, the use of projects is likely to spread. Several pioneers noted that while projects might not be essential to a modern calculus course, their use is one of several options, such as writing and group work that students should experience.

Writing about mathematics is an integral part of student projects, but it is also a pedagogical tool used by many instructors in a variety of other settings. "Until people put things into words, they don't really understand." "Clear thinking can't be divorced from clear writing." "Writing determines understanding." "Writing helps [students to] understand concepts." "Communication skills in general are part of the learning process." "Training them to write is tough," and grading writing may be difficult, but "calculus reform benefitted from the timing of the writing across the curriculum movement." Students may be asked to answer questions with full sentences that require them to say what they mean. Summaries of discussions, presentations of ideas in students' own words, and journal entries are other options in the writing process. "Students will have to learn to write," "to look at a real situation and describe it verbally and mathematically."

In addition to being one of the multiple ways to represent concepts and ideas, student writing is a valuable way to find out what students know. Software tools are making it easier for students to write about the mathematics they are doing, to set out convincing arguments integrating verbal, graphical, analytical, and numerical components, and to take pride in their finished work.

While neither the Tulane conference nor the *Calculus for a New Century* colloquium set forth a prescribed new syllabus for calculus, many people thought calculus reform would be centered on **content changes**. In fact, only modest changes have been made in the topics covered in calculus courses. The most significant changes are the early appearance of differential equations and the pruning of the integration techniques thicket. Some reform projects did begin by examining the course content and investigating what was actually used in other courses in mathematics and the client disciplines. However, no topic should stay in a course just because it is used later. Each course should be a coherent entity. For many pioneers, content changes are a continuing evolution as courses focus on the big ideas of differentiation and integration. "We should give greater visibility to the main ideas, less to subservient cases." "A lot of routines should go unless they develop concepts." "Focus on fundamental ideas students can appreciate now." "Other objectives, especially learning to learn," are causing content changes, too. When the question changes from "how long does it take to say this in class," to "how long does it take the student to learn this," changes must be made. As described by one pioneer, the result is "a more horizontal curriculum than the old rigid vertical one; filtering [out of students] is reinforced by the vertical curriculum."

Other pioneers see the changes to be in the "packaging" and "emphasis" with real change yet to come or not of great importance. "Calc I is stable, an intellectual whole through the Fundamental Theorem." "All mathematics is exciting and fun." "The change is not in content, but in the way it is taught." "Students need to understand the language [of calculus] and what the mathematics says."

The opinions of those interviewed span the full range of views on the role of **concept development** in a modern calculus course. "People were always doing [it]; it's necessary, but not reform." Concept development has "increased by a factor of 2000." "We need to do much more with it." These three quotes may in fact all be correct; we always thought we were developing the concepts of derivative and integral, but now we know much more about what students actually learn of the fundamentals of calculus in a traditional course. "This is the raison d'etre of the last ten years." "It's done better now than before; we have a better idea of what the big ideas are." The use of multiple representations and technology were both mentioned repeatedly as extremely useful in helping students develop "understanding in a rich and varied environment." "Technology can really help students play with ideas." This may mean that in many modern calculus courses, concepts take precedence over such things as formal proofs. "It takes time and energy [to develop conceptual understanding]; we pay with less computational work." "Multiple representations do work."

The full potential of technology for building conceptual understanding has yet to be used according to several pioneers. "People are not using technology to teach in different ways." "Better use of technology should be the main part [of change]." Besides using technology better for concept development, we still need to know how to do it better in general.

Despite the oft-repeated characterization of calculus reform proponents as being opposed to lecturing, committed to using computers, and dedicated to collaborative learning, the reality about **pedagogy** is quite different. "The only requirement is that the professor have real contact with the students and build on [his or her] individual strengths." "It's a matter of personal tastes and institutional mores." "We must be careful about orthodoxies. The lecture method can be done well." "Whatever works." "Find what pieces will work for you." As a number of the pioneers said, the point is to "change the focus to student learning." "Most mathematicians could profit by thinking more about learning than teaching." "Sympathetic teachers, in a false attempt to aid students, changed problems into templates. A really sympathetic teacher will force students to think." It may be comfortable to teach as we were taught, but few of our students will be mathematics majors, and other methods may be more successful for all students.

Related to changes in pedagogy are changes in assessment techniques. One-hour, in-class, short-answer tests provide only limited information about student learning. Using a variety of activities and assignments in a modern calculus course requires a variety of assessment tools. We can learn much from other disciplines about evaluating projects, writing, and collaborative work.

Success in mathematics depends on habits of mind and ways of learning peculiar to the discipline. **Research in how people learn mathematics** is beginning to inform classroom practices, but many remain skeptical of how much educational research will contribute. Skepticism arises in part from mathematicians' experiences with a very different type of research. However, as mathematicians work to become better teachers, the expectations for educational research grow. "We need research on understanding and concept development." "We've been making random discoveries; we need systematic study." "To achieve the goals of calculus reform, we need research to support continued progress." "We need more, but we can't wait until the results are in." "We need research to justify some of our articles of faith; e.g., communication, group work." Although some pioneers say research had "a major impact on how courses developed," others would agree that "research results are not in a form useful to those in the trenches. Informal results still predominate." The consensus of the pioneers is that whether significant results are available yet or not, research into how students learn mathematics is important and needs to be undertaken by more people. The fraction of mathematics teachers doing educational research will always be small, but "thinking about calculus is important" for all of us. If a goal is "students having robust understanding," then we must find "instructional strategies and environments to accomplish this."

Roots of Reform

Research is but one component of the continuing development of modern calculus courses. In the last ten years, though, tremendous amounts of time, energy, and creativity have already been devoted to crafting and teaching calculus courses. According to *Assessing Calculus Reform Efforts* (ACRE report, MAA 1995), one-third of all calculus students are in courses that have been affected by the calculus reform movement. These courses are at two-thirds of the colleges and universities. One might ask why and how this happened. The quick, and somewhat cynical, response is money, namely funds from the National Science Foundation's Calculus Curriculum Development Program. Certainly funding played a significant role, but by itself it would not have been sufficient to produce changes of the magnitude we have seen. The commitment of the NSF to changing calculus instruction "legitimized," "gave prominence," and "respect" to calculus reform. Equally important was the Foundation's support of implementation as well as curriculum development and the "broader effort to get results out and used." "NSF's biggest contribution was not money, but identification of individual projects."

Another major factor driving efforts to change calculus instruction was timing. Dissatisfaction with calculus courses coincided with a demographic bump in the professorate. Pioneers described "wide-spread dissatisfaction with the calculus program at the grassroots level," "faculty who had 20–30 years of struggling to make calculus exciting," and a course that was "wrong." "Even defenders [of the traditional course] were frustrated with the results." "The sterility bored us." "People are desperate; there's got to be a better way, livelier, not lobotomized." Having taught for twenty plus years, many faculty members were confident "we can't do worse." Only half-jokingly, one interviewee suggested that calculus reform caught on because of "a middle-aged, intellectual crisis." Perhaps. Other trends also made the timing right for change. "The world is changing so not just calculus is changing. There are similar changes in other disciplines." "There's been a change in our role as university faculty, in our professional and societal responsibilities." "University teaching had been stagnant for 30 years." "We owe more to students and society." According to the pioneers, developments in technology such as graphing calculators and PC's played only a minor role in the timing of calculus reform.

The way in which discussion of calculus reform was launched was a factor in the spread of change. "A group talking publicly about calculus made it legitimate to question what was going on in calculus programs." "Calculus reform was given a good name by well-respected mathematicians." "Something here appeals to the mathematics community. It's open to questions about goals, about what students are doing." It certainly helped that calculus reform was "an opportunity for mathematicians to play a leadership role in the university and in education."

Loud and clear is the message that the deepest, strongest reason for the existence of modern calculus courses is that they are exciting and fun. "It's been fun for a lot of people." "People are pushing, because it's fun; it was less and less fun in the old version." The new courses are "more satisfying for instructors; this is a payoff they didn't expect." "You see the way students respond, and you and the students get excited. It's fun and intellectually engaging." "It has opened lots of doors offering faculty a chance to renew as teachers."

"Reform may not have found **the** answer; we're on a road with a long way to go. We must keep examining how and why."

I want to express my sincere appreciation, on behalf of the other editors and myself, to those who were so generous with their time and thoughts. Those interviewed are listed below. This essay is a pale reproduction of the rich, passionate, thoughtful, and perceptive conversations that I enjoyed with these people. I wish you, the reader, could also have had that marvelous experience.

Those interviewed: James J. Ca (Smith C lege), Raymond J. Cannon, Jr. (Baylor University), William J. Davis (The Ohio State University), Thomas P. Dick (Oregon State University), Ed Dubinsky (Purdue University), Deborah Hughes Hallett (Harvard University), William E. Haver (Virginia Commonwealth University), Stephen R. Hilbert (Ithaca College), Elgin H. Johnston (Iowa State University), John W. Kenelly (Clemson University), Arnold M. Ostebee (St. Olaf College), David John Pengelley (New Mexico State University), A. Wayne Roberts (Macalester College), Donald B. Small (United States Military Academy), David A. Smith (Duke University), Gilbert Strang (Massachusetts Institute of Technology), Keith D. Stroyan (University of Iowa), Thomas W. Tucker (Colgate University), J. Jerry Uhl, Jr. (University of Illinois), and Franklin A. Wattenberg (Weber State University).

Nonalgebraic Approaches to Calculus

Thomas W. Tucker
COLGATE UNIVERSITY

The 1986 Tulane Conference on Calculus was prescient in a number of remarkable ways. Its emphasis on pedagogy, as at least as big an issue as content, has been a guiding principle for calculus reform, and may be the reason calculus reform has been so widespread, influential, and persistent. The recommendations from the Content Workshop Report at that Conference (pages vii–xiv of *Toward A Lean and Lively Calculus*, MAA Notes No. 6) have also found wide acceptance. In particular, the graphical and numerical approaches urged by that report have become the cornerstones of at least two of the most successful calculus reform texts, Ostebee/Zorn and Hughes–Hallett, *et al.* Just as calculus reform is unlocking the strangle-hold of the lecture method on pedagogy, it is also unlocking the strangle-hold of algebra on content.

An Example With Solutions

Let us begin, as always is best, with an example. A car going 50 feet per second puts on its brakes and comes to a stop in 5 seconds. How far does it travel?

Like most questions, this one is ill-posed, although it should be pretty clear how to figure it out. Is the acceleration constant? Yes, because the question couldn't be answered otherwise. Also, a constant brake force is a reasonable assumption. Over what time period are we computing distance traveled? It must be from the time the brakes are first applied to the time the car comes to a stop.

Here are some solutions:

Solution A: The acceleration must be -10 ft/sec/sec to slow the car from 50 ft/sec to 0 ft/sec in 5 seconds. We then have $a = dv/dt = -10$, $v = -10t + C$.

Since $v = 50$, when $t = 0$, we must have $C = 50$: $v = ds/dt = 50 - 10t$, $s = 50t - 5t^2 + C$.

We assume the position $s = 0$ when $t = 0$ so $C = 0$: $s = 50t - 5t^2$.

Then plug in $t = 5$ to get the distance traveled: $s = 50(5) - 5(25) = 125$ ft.

Solution A': $a = dv/dt = 10$, so $v = ds/dt = 10t$ so $s = 5t^2$. When $t = 5$, $s = 5(25) = 125$ ft.

Solution G: The graph of velocity versus time is a straight line going from $t = 0$, $v = 50$ to $t = 5$, $v = 0$. The area under the graph of v is distance traveled, and the area is a triangle with base 5 seconds and height 50 ft/sec. Thus the distance traveled is $\frac{1}{2}bh = \frac{1}{2}(5)(50) = 125$ ft.

Solution N: The velocity at one second time intervals is given in the table below:

t	0	1	2	3	4	5
v	50	40	30	20	10	0

During the first second the car travels somewhere between 50 and 40 feet, the second second somewhere between 40 and 30 feet, and so on. Upper- and lower-bounds for the distances traveled are

$$50 + 40 + 30 + 20 + 10 = 150 \text{ ft.}$$
$$40 + 30 + 20 + 10 + 0 = 100 \text{ ft.}$$

Splitting the difference gives 125 ft.

Solution V: The velocity goes from 50 to 0 ft/sec so the average velocity is 25 ft/sec. Thus the distance traveled is (25 ft/sec)(5 sec) = 125 ft.

I would characterize these solutions, in order, as algebraic (A and A'), graphical (G), numerical (N), and verbal (V). Each has its strengths and weaknesses.

The Algebraic Solution

The algebraic approach is probably what most teachers and students expect to see in a calculus course. After all, as one of my colleagues likes to remind me, calculus was developed so that problems much more complex than this can be handled simply by algebraic manipulation. To paraphrase G.H. Whitehead, the advance of civilization is measured by the number of things one can do without thinking, and calculus allows a lot of people to do some very sophisticated things (say, to the ancient Greeks) just by turning a crank. Archimedes would be astounded to see the area and volume problems that are now solved routinely by hundreds of thousands of teen-agers. To be honest, I am astounded, too, if I stop to think about it. Although mathematics teachers are always wringing their hands about some students' abusive treatment of algebra, I have read enough essays and school newspapers to know that the algebraic skills of my calculus students far exceed their writing skills. They have had an awful lot of algebraic training, and it shows. Indeed, calculus as a *tour-de-force* of algebraic manipulation is perhaps the appropriate capstone to a mathematical education that is overwhelming algebraic in viewpoint.

A lot can go wrong with algebra, however. Its power is dangerous, and putting it in the hands of some students is like putting a Porsche in the hands of a teenage driver. Some pretty horrible things can happen. Solution A' is one example. When I ask this question on a test, I see Solution A' almost as often as Solution A, and I get tired of explaining why the correct answer of 125 feet does not get full credit. Solution A' reveals part of the problem with the algebraic approach: the actual manipulation is meaningless. If students can do a problem without thinking by turning a crank, how does one explain to them that they turned the crank incorrectly, or that they turned the wrong crank entirely? Why should students care how well they turn the crank? It is not exactly something to take pride in. That is why students are blasé about their algebraic mistakes in the same way mathematicians are blasé about their own arithmetic mistakes.

Another problem with algebraic knowledge is that it does not stick. Students flush formulas as soon as the exam is over. Even my best students can't tell me the antiderivative of $1/(1+x^2)$ when they are more than a year away from calculus. In fact, it has slowly dawned on me that the only reason I remember a lot of calculus formulas is that I teach them every year. Unfortunately, when students lose their algebraic knowledge, they are often helpless to reconstruct it or replace it with something else. The same students, who would astound Archimedes with their ability to compute the volume of some exotic solid of revolution, would appall him when they ask for the formula for the volume of a cylinder needed for some max/min problem.

Perhaps the most glaring trouble with algebraic representation is how rarely it is used outside the sciences. People do not communicate ideas algebraically in everyday life. This is partly because most everyday phenomena do not have nice algebraic representations, but it runs much deeper than that. Even when there is a reasonable algebraic representation, for example, to compute line 41 on a tax form, people use words or tables to describe the situation, not algebraic symbols. If a boss asks for an analysis of some problem, she expects to see graphs, charts, tables, and one-page verbal summaries; the last thing she wants is algebraic formulas. Since computers so easily generate striking graphical and tabular displays of data, even scientists rely less now on algebraic representation as the primary means of conveying information. There is a comedy line that one of the secrets about being an adult is that you never need algebra. Like it or not, there is more than a little bit of truth in this line. Mathematicians would do well to heed the public's viewpoint that much of mathematical education is irrelevant, especially when it comes to algebra.

The Graphical Solution

What about the other solutions to the car problem? The graphical solution is certainly simple and elegant. It does depend on the rubric that distance traveled equals area under the velocity curve. Although one might argue that this is just another example of a formula learned by rote, one can use the units measured along the time and velocity axes to explain why the area of each rectangle in the graphing grid is measured in feet and hence represents distance. In my experience, I find students have difficulty seeing the area under a curve as representing a physical quantity, but this viewpoint can be taught and it does stick. In fact, I believe students can be taught to manipulate graphs just as easily as they manipulate symbols. The one question on my tests, that I can now be sure every student gets right, is sketching the graph of the derivative given the graph of a function, or *vice versa*. It shouldn't be surprising that students can handle a purely graphical approach.

After all, economics courses have always argued from graphs alone, usually out of necessity since algebraic representations may not exist. I doubt any mathematicians feel the students in Economics I are any better than those in Calculus I; if the Economics Department can teach students to analyze problems graphically without algebra, so can the Mathematics Department.

Graphical knowledge is robust and can be reconstructed many years after the fact. I have asked my wife, who teaches computer science, the car question and suggested trying either an algebraic approach or graphical approach. It has been more than 25 years since she took calculus and she immediately told me she couldn't do it algebraically. On the other hand, she drew the velocity graph and saw that solution in an instant. Visual images are strong, and, although they are hard to quantify, they can carry deep understanding. A picture is worth a hundred symbols.

Of course, one obvious advantage of the algebraic approach is that it is widely applicable. If the acceleration were not constant, the same steps of Solution A could still be followed to get the distance traveled. The trouble is that one would still need an algebraic representation of the acceleration and that may not be realistic. The graphical approach is also widely applicable, especially where the information about the given acceleration is qualitative. Consider the following variation of the car problem. Suppose the car slows down from 50 ft/sec to 0 ft/sec in 5 seconds so that the acceleration function is continuous and symmetric about the time $t = 2.5$. In other words, the brake pedal is slowly depressed from $t = 0$ to $t = 2.5$, and then released in the same manner from $t = 2.5$ to $t = 5$. What is the distance traveled?

The area under the acceleration curve is -25 ft/sec from $t = 0$ to $t = 2.5$ and -25 ft/sec from $t = 2.5$ to $t = 5$. By symmetry, the velocity curve will be symmetric about the point $t = 2.5$, $v = 25$. Again by symmetry, the area under the velocity curve is the same as the area under the horizontal line $v = 25$. Therefore the total distance traveled is still 125 ft.

A graphical approach could handle other variations as well, such as comparing the distance traveled when the braking force is applied more at the beginning or more at the end. Just as the algebraic approach can be applied to variations in the algebraic representation of the acceleration, the graphical approach can be applied to variations in the graphical representation of the acceleration.

The Numerical Solution

To some minds this is not a solution at all, since the final answer is an unjustified average of an upper- and lower-bound for the distance traveled. Of course, the approach works not just because of blind luck, but also because averaging is the correct thing to do for a linear function (and since most functions are locally linear, it is also a good idea for arbitrary functions). Moreover, there are a number of advantages to this approach. One is obvious: this method takes seriously the challenge of the problem to find a numerical value for the distance traveled. The numerical approach provides not only an answer, but also an upper- and lower-bound for the answer, and hence an immediate verification of the reasonableness of the answer as a number.

Using a table of values is so easy and natural that students often feel it is illegal. Even people who should know better feel that way. A mathematician once told me of sitting next to an engineer on a plane and of being asked, as a mathematician, how to calculate the value of a certain definite interval the engineer had encountered. The mathematician asked, "Well, what did you try to do?" The engineer replied, "I couldn't remember any of my calculus formulas, so I divided the interval in pieces ..." The engineer proceeded to describe his own self-discovered version of the trapezoid rule. The mathematician asked how well the method worked. The engineer replied "Just fine. But how should I have done it?" The mathematician responded, I am happy to say, "Exactly the way you did!"

Mathematicians under-estimate the public's interest and facility with numbers. Although we all know the horror stories of store clerks with calculators, and newscasters mistaking million for billion without blinking an eye, most people treat numbers with more reverence and ingenuity than they treat words. If you don't believe me, read the sports page. Tables of values appear everywhere except in calculus textbooks. What really made the business community pick up on personal computers was word-processing and spreadsheets. Somehow mathematics education, at the calculus level, has failed to exploit this natural interest in numbers. Students are not encouraged to experiment numerically to find a limit, say for an asymptote, just by plugging numbers into a function and seeing what happens. They are not taught to recognize a linear or exponential function from a table of values. In fact, a lot of mathematicians might not immediately recognize a linear function of two variables just from a table of values for the function (what should the rows and columns look like?).

Beside being easy and natural, a numerical approach is also powerful because it is so widely applicable. If the acceleration for the car problem is changed, there is no guarantee that it can be antidifferentiated in elementary terms, and hence the algebraic approach may fail. On the other hand, a numerical approach via tables of values and Riemann sums is reasonably insensitive to the form of the acceleration. Moreover, the underlying idea of subdivision and summing applies, as we know, to a myriad of situations other than acceleration-velocity-position.

The Verbal Solution

Since the original question is presented verbally, not algebraically or graphically or numerically, all the solutions are to some extent verbal. The last solution, however, depends on the connotation of the word "average" and uses no algebra, graphs, or tables, so it is the most verbal of the four solutions. It is also the most suspect and probably least accepted by mathematicians. We use algebra precisely to avoid the vagueness of words. Nevertheless, the given solution is precise. Luckily, the term "average" has been given a technical definition by mathematicians that coincides with the common sense notion of average, especially for linear functions. As a consequence, the rule that "distance traveled is average velocity times time" becomes just a restatement of "distance traveled is the definite integral of velocity," when the term "average" is interpreted in terms of a definite integral.

Moreover, this approach can be used for many other applications of the definite integral. For example, how much work does it take to pull up a 100 foot chain dangling from the top of a building if the chain weighs 1 pound per foot? Work is force times distance. As the chain is pulled up the building, the force applied varies from 100 pounds to 0 pounds. The average force is 50 pounds, so the work is (50)(100) = 5000 foot pounds.

The real reason, however, that the verbal approach is so important is that people communicate ideas verbally. Other disciplines have long known that to write is to think. Ideas only take shape when they are put into words, sentences, and paragraphs. One common theme of calculus reform courses is that students should write more: projects, laboratory reports, essay questions on exams, instructions to "explain your answer." The failure of standard calculus courses to come to grips the importance of verbal communication of mathematical ideas comes home to roost in later mathematics courses where students first have to write proofs. One reason students find abstract linear algebra such a culture shock is that the proofs are mostly verbal, and the calculus courses these students have taken up to this point have not required them to write clearly and precisely. In part, the full import of Solution V might be just as difficult for students to understand as a proof that the columns of an invertible square matrix are linearly independent.

Conclusion

Calculus has been taught from an algebraic viewpoint for so long it is difficult for mathematicians to see it any other way. The result is a course that is isolated from the disciplines it serves, including mathematics itself. The kind of technical, context-free, algebraic manipulation so prevalent in a calculus course simply does not occur in chemistry, biology, economics, or even physics. In fact, it does not occur either in linear algebra, real analysis, abstract algebra, probability, topology, geometry, or much of the undergraduate mathematics curriculum. If anything, the algebraic emphasis of a standard calculus course is really anachronistic. Algebra was used for the three hundred years since Newton's *Principia* partly because numerical methods were inaccessible to humans before the invention of the computer. The computer not only allows direct numerical and graphical solutions to real problems, it is also now good enough and easy enough to use for symbolic manipulation, that skill at algebraic manipulation by hand may soon become no more useful than skill at long division by hand.

Nonalgebraic approaches should be a defining characteristic of a contemporary calculus course. The car example discussed in this paper shows how these different approaches handle the same problem. The trouble is that most calculus students, if given this problem, would head for Solution A, or unfortunately A'; they would immediately try to fill a blank page with symbols and equations. I look forward to the day when their instinct is to draw a graph, or construct a table of values, or even write a little essay.

Towards Active Processes for Teaching and Learning

Mai Gehrke and *David Pengelley*
NEW MEXICO STATE UNIVERSITY

What are the real goals of 'reforming' calculus teaching? Some commonly accepted aims are making calculus more relevant and understandable to students, making it 'lean and lively,' and having students acquire and meld the tools of calculus to solve multi-step or open-ended problems. However, all these aims are actually just part of an overarching goal: having students actively involved and taking initiative in their own learning, in fact learning how to learn for themselves. Our students should take responsibility for and charge of their own learning, developing their own process for becoming independent learners, and thus end up with a sense of personal ownership of the results of their labor. The commonly accepted aims listed above are good first steps on this road towards helping students become active learners: by creating material relevant to students' curiosity we make it possible and attractive for them to take interest; by making the syllabus lean enough, we allow time for the high level of absorption inherent in active learning; by having students solve larger contextual problems rather than template problems floating in a vacuum, calculus becomes understandable and useful in a more real way.

So what are the means for achieving the broad goal of making students active learners? Is it accomplished by incorporating technology, or by adopting a 'reformed' text, or through substantial individual or group projects? All these can be useful pedagogical tools in trying to reach the broad goal. However, if we lose sight of the overarching goal, assuming that adopting one or more of these specific tools is the essence of reform, then we will fail; when the tools themselves become the focus, we depart from the necessary primary emphasis on the active involvement process. Another reason not to restrict one's view to a specific tool is that if our students are truly becoming active learners, due to their individuality they will each develop personal learning processes which thrive on different tools.

Since our ultimate goal for students is a focus on their learning process, the tools we use should always be treated merely as objects within that process. Their importance and usefulness should be kept in perspective, and thus the teacher's focus should also shift away from objects. Objects such as text materials, computers, labs, lectures, reading, writing, projects, group work, exams, and portfolios are colors on the palette from which teacher and students can choose during the ongoing process of each student's learning. Ultimately, we believe this shift will need to encompass all levels of the educational structure: students, teachers, and the educational community as a whole.

Our professional community has the same responsibility to teachers that teachers have to students: to create an environment in which teachers will naturally evolve an ongoing active and creative teaching process. In particular, teachers must be encouraged to show individual initiative in order for reform to succeed. Each teacher must 'make it their own,' selecting from a community palette of ideas and resources.

Our own experience is that in a department where collaboration in teaching innovation is strongly encouraged, while individual instructors still make their own decisions about how to teach their classes, we have an unusually high level of participation in 'reform' without forcing involvement. Our department values the varied contributions of many teachers, and these are continually being synthesized into our own current version of reform; individuals and groups of teachers communicate their fresh approaches, often resulting in the incorporation of their ideas into the reform undertaken by others. What our department 'lacks' is an orthodoxy of reformed teaching materials or other objects, since we recognize that individual teachers will go about creating process differently.

Giving teachers the freedom to develop their own personalized reform is what incites initiative from them, and thus ultimately from their students. Faculty can then become active teachers, and the multitude of approaches they develop will naturally induce healthy cross-fertilization. Of course change in a given department may not start with individual initiative from each teacher, but can nevertheless evolve into an organic environment for reform provided the focus of change is not so rigidly tied to certain objects of reform that it frustrates the development of individual process for teachers.

Thus the community should nurture a fluid atmosphere, in which adopting pre-existing reform materials can go hand in hand with initiating an individual teacher's process of change, and should provide a library of resources from which teachers can create personalized pedagogical tools for their own teaching. With this kind of individual growth, change will happen more slowly than with a superficial imposition or adoption of prepackaged reform, but will surely be more longlasting, and more faithful to the goals above; a teacher creating her/his own materials, or modifying those created by others, will be an active teacher, whereas simply using prepackaged materials will not stimulate a teacher to emerge from passivity.

What we have experienced at New Mexico State University gives an example of how this can happen on a sizeable scale. We began with a small group of faculty initiating change. Then a larger group of faculty found themselves enticed into getting involved, and this subsequently shifted the pedagogical nature of our efforts. In this way the group of faculty involved has kept growing, and our direction has evolved with this growth. Thus our 'reform' has happened in stages, each reflecting a new horizon which only became visible at the end of a previous stage. Our evolution can provide an example of the dynamics of individual and departmental cultural change.

For us, change began in 1987 as a response to unsatisfactory student performance in calculus. Two faculty members came up with some basic ideas for improving the situation. One was to grab students' attention—we were forever hearing from students that they could not work on their mathematics homework because they had assignments due in other classes. A second idea was to have students do some real mathematics—problems that they would solve and explain as opposed to ones to which they merely supplied an answer. Basically, we wanted them to think and we needed a way to encourage them to do that.

Projects were designed to get students to think for themselves on major multistep, take-home problems, working individually or in groups. We hoped to alter fundamentally students' view of what mathematics is all about and simultaneously build their self-confidence in what they could achieve through imaginative, theoretical thinking. The projects resemble mini-research problems. Most of them require creative thought and all of them engage students' analytic and intuitive faculties, often weaving together ideas from many parts of calculus. While many of the projects are couched in seemingly real-world settings, often with engaging story lines, they are all in a sense theoretical. One cannot do them without an appreciation of the ideas behind the method. Students must decide what the problem is about, what tools from the calculus they will use to solve it, find a strategy for its solution, and present their findings in a written report. This approach yields an amazing level of sincere questioning, energetic research, dogged persistence, and conscientious communication from students. Moreover, our own opinions of our students' capabilities skyrocketed as they rose to the challenges presented by these projects, and some other faculty and graduate teaching assistants were smitten and wanted to get involved.

Even though the idea of having students work on projects seemed a revolutionary idea at its conception, it was a small enough step that a number of faculty felt comfortable about incorporating a project or two in their courses. The new teachers wanted to create their own projects, or modify old ones, each bringing a unique perspective to what a 'project' should be, and thus they became active in reform. The motivations and types of projects written by this conglomeration of people varied and added breadth and scope to the nature and efficacy of using projects in teaching. Over 100 projects were developed by five faculty during this period, and published in the MAA book *Student Research Projects in Calculus*, along with several chapters detailing the logistics of assigning projects and advice for instructors.

In 1990 the program expanded and branched in various directions. Numerous other faculty in the department volunteered to use calculus projects in their classes, and we began the development of a discovery-project based vector calculus and differential equations curriculum (in which a continuous sequence of discovery projects forms the context for learning all the material of the course); we also started a collaborative program with local high school teachers to bring projects into high school mathematics courses.

As new faculty became involved in teaching with projects, they injected fresh ideas into the program and the projects approach itself evolved. Although introducing projects was a valid first step, we realized this had created somewhat schizophrenic courses in which students worked on projects outside class, while the classroom continued to function in a traditional style. Even though we felt that the activities involved in working on projects were effective in stimulating students to think and to learn mathematics, our day-to-day classroom activities remained largely unchanged. This provoked a new stage in our development.

Thus in 1991 a group of faculty pioneered a major new emphasis on cooperative self-learning both in and out of the classroom, developing structured in-class assignments called 'themes.' A distinct change is that themes are used to introduce the core material of the course and much class time is spent working on them, with less time on lecture, whereas the projects were completed outside class and contained material over and above day-to-day course work.

In a theme assignment, students learn and write about core course material while working in groups with the instructor serving as a resource. When themes were first assigned, students completed a written theme report every week. Experience has tempered this pace somewhat, and we are now assigning three to six themes per semester. Today, several instructors are blending the theme approach with the discovery-project methods developed in vector calculus and differential equations courses. Other ideas, such as class discussions, student presentations, and mastery skills exams, are being tried also. These somewhat independent directions seem to be cross-fertilizing each other's growth.

In retrospect, we see that each of the tools we develop leads to new pedagogical challenges. For instance, we were pleased with the high level of student initiative and achievement that projects elicited, but we wanted to get away from the passive role of our students during a lecture. With themes and writing assignments, students were active in the classroom, but we realized the teacher should be more than just a resource for individual students or small groups to call on; in fact, this placed the teacher in too passive a role. The teacher should be providing leadership to the class as a whole, in order to take advantage of having all the students and teacher together. Sometimes while students were working on themes, most groups would generate a common question, which naturally led to a whole-class discussion moderated and guided by the instructor. In fact the guided class discussions based on students' questions arising from their active work emerged as one of the most successful and productive aspects of this student-centered classroom. We now view such guided class discussions as an important tool in their own right, and we have found other student activities which benefit from and enable these discussions. For instance, another of our aims has been for our students to become capable and active readers of mathematics. This requires breaking the vicious cycle in which instructors lecture text material to students because they know students don't actually learn it from reading, and students have little incentive to read because they know their instructor will lecture it to them. We have found that if we demand students read in advance, and write commentary and questions about their reading, then these questions can form the basis for active class discussion, bypassing the vicious cycle and leading to more productive and satisfying classroom learning.

Theme assignments have also prompted us to incorporate structured means of improving student skills at mathematical and prose report writing. After incorporating handouts on writing, and learning how to guide students in honing their writing skills, we have seen an incredible improvement in their ability to write. Reading and writing in mathematics have emerged as important features of reform at both the undergraduate and graduate level, and can be viewed as a new stage which has spread far beyond our calculus courses. These new emphases have merged with innovative efforts of other faculty who were never even involved in our calculus reform program.

Of course, our means and methods for grading have also changed drastically over this period of reform. When we introduced projects and themes as learning tools, we also used them as an important means of evaluation. In comparison we found that traditional exams have little to do with learning, and we now primarily value means of evaluation that are also learning tools. Our methods of grading began to change as well, since already the projects required us to learn how to evaluate written reports and group work. We came to realize that detailed numerical grading was poorly suited to grading large written reports, so we have been learning how to evaluate student work in a more holistic fashion. A benefit of qualitative holistic grading is that students get specific feedback on how to improve their written work. At first we worried that students would feel uncomfortable not having points attached to every aspect of their work, but we found that they readily accept and appreciate qualitative feedback and evaluation; it is easier for them to see the qualitative nature of the distinction between A and B work, provided the criteria are clearly explained, than to understand the difference between grades of 89 and 90.

After seeing the benefits of holistic evaluation for individual assignments, it is natural to consider extending this methodology to evaluation of the totality of a student's work. Recently this has led some of us to a portfolio approach, in which the student prepares a showcase of their entire work for the semester, and this portfolio is evaluated as a whole at the end. In contrast, traditional grading is by nature fragmented, encouraging a disconnected view in the student of both the course and their own work. By assuming responsibility for collecting, organizing, and presenting all their course work in a portfolio, students become aware of the big picture in both the subject matter and their own performance.

Looking back on all these changes, we see that they involved a relinquishing of total control. While this can be a frightening prospect, it is necessary if students are to assume more responsibility and control of their own learning. Fortunately, if the balance of control is shifted gradually from teacher to students, through a slow process of evolution, total loss of control may be avoided. The reward is the opening of new vistas for both teaching and learning, in which the instructor becomes an expert guide, facilitator, and coordinator. Even though the original purpose is to improve student learning, there is a tremendous revitalizing benefit for teachers, as our interaction with students and colleagues becomes more rewarding, and the results of our efforts become more meaningful.

Each of the specific changes and pedagogical discoveries that we have made along the way has in a sense forced itself upon us as an inescapable outgrowth of a previous change. This process, and the collegial atmosphere that has made it possible, are in our minds the essential features of reform as we have experienced it. Alan Schoenfeld, in the preface to the recently published book *Mathematical Thinking and Problem Solving*, referred to our initial seed, namely student projects, as a Trojan Mouse, and that is truly what it has been, subversively driving the scope of change far beyond what we could originally imagine. The other essential feature of our reform is the atmosphere of faculty collaboration in teaching innovation; it has nurtured almost everything we have accomplished, and in a way which has fostered individual faculty ownership of both the process and the results, anchoring it deeply in the fabric of our department community.

Technology in Mathematics Instruction

John Kenelly
CLEMSON UNIVERSITY

Everyone bench marks the calculus reform movement as starting with the 1985 Anaheim Panel "Calculus Instruction, Crucial but Ailing" and the 1986 "Lean and Lively" Sloan Conference report. Even though I participated in both, I did not return to my calculus classroom with a crusading zeal to deliver "calculus reform." In all honesty, I tried to implement many of the suggestions but, in fact, I really made just a few changes that blended well with the mandated department syllabus. Maybe more improvements would have been added as the years progressed, but my personal "jump discontinuity" was a classroom set of HP–28C graphic calculators in the fall of 1987! I told the department that someone *had* to try this fascinating new gadget in the classroom, but the instructor would need the freedom to experiment with the course. Like a fool I accepted the assignment, but there was a stipulation that I would have to carry the experimental group through the full course sequence. This meant that I had to worry about attrition for the first time; otherwise the enrollment would fall below our teaching credit level. The experiment was highly visible, so I could not get by with easy grades. This meant that I had to examine each course topic, in the spirit of the Sloan Conference, but deliver them with technology that no one had dreamed about at Tulane. In that important meeting in January 1986, we thought that the "solve" key and the "numerical integral" key were ready to make changes in calculus instruction, but we were all months away from seeing a graphics calculator, much less one with internal symbolic manipulating capabilities. The end result was that everyone actively participated in class. My students discovered each other and I rediscovered them. I now worried about how they learned as well as what they learned. Mechanical manipulations lost their importance and concepts became everyone's concern. When the students saw the machine "do calculus" they did not need to be told that they needed higher goals. With the new focus, the most noticeable change was the total absence of the question, "Where will I ever need this?"

Technology is changing the way we teach. Not because *it's here*, but because *it's everywhere*. Life itself revolves around electronics and we are in the information age. Today's automobiles have more computing devices than the Apollo capsule. Business is a vast network of word processors and spreadsheets. Engineering and Industry are a maze of workstations and automated controls. Our students will have vastly different careers and we, the earlier generation, must radically change the way that education prepares a significantly larger part of the population for information intensive professional lives.

Mathematics instruction must change, and one of the most striking changes is the need for a broad section of the population to understand and use calculus concepts. This historical threshold to science and engineering careers is now overwhelmingly the intellectual base of the information age. How is that challenge going to be addressed? Technology *has* to help and that is the focus of this paper. Since 1987, the author has required *every* student in *every* class to have a graphics calculator, and that is the background that he brings to this article.

General Observations

The calculator vs. computer debate is on the wrong track and particular capabilities arguments are pointless. Hand-held devices are advancing faster than the argument about specific features can be settled, i.e., both devices will have the feature before you finish the debate. We should look in another direction and focus on *activities*. That is, what activities are best performed on portable devices and which ones are more appropriate at computer stations? To this discussion, the author would suggest the following:

Portable Device	Computing Station
class testing	national standardized testing
day-to-day study	major projects
casual explorations	large scale simulations
moments of opportunity	formal demonstrations
personal notations	data bases

As you see, the debate should be on the nature of the activity and not on computing capabilities. When you look at computing devices in that perspective, you note that technology's role has never been a question of *whether computers or calculators*, but where *each best fits*. Accordingly they both have a critical role and greatly complement each other.

Personal is the key word for hand-held devices, and *communications* seems to be the key for computing stations. Without a doubt, computing stations will never have the privacy sense that calculators provide, and I think that has very important gender implications in mathematics instruction. Most of the folklore is anecdotal, but too many colleagues have reported that computer requirements compound the problems of attracting females to mathematics. And, at the same time, calculator users report positive gender experiences. This is a problem that demands study. There does not seem to be the same consistency in the reports on the effects on minority students, but this is an equally important task for the learning theory researchers.

Testing is the reality check on instruction, and technology is making major in-roads into the field. Computer Adaptive Testing, i.e., computer branching on candidate responses, is based on the new measurement field of *item response theory* and computers are essential. We already see CAT being used in the new computer delivered Graduate Record Examination and the National Teacher's Examination, and there are scores more of CAT-based tests coming. CAT generates a unique sequence of questions for each candidate, so that frees everyone from schedule constraints. They impose on the candidate a smaller individually focused set of questions, so they are very efficient.

The national testing programs are having major influences on technology in mathematics instruction. To the author and many others, allowing calculators on the Scholastic Assessment Test (SAT) and requiring graphic calculators on the Advanced Placement (AP) examination were major improvements that significantly changed mathematics instruction in the schools. In the first case, routine arithmetic calculations are properly de-emphasized and, in the second case, AP Calculus study moved into the technology age. The national AP graphing calculator requirement appears to contradict the author's position in the above chart. It does *not* because, I personally feel that in the future, the AP test will follow the national testing trends and move into a CAT format.

There is no doubt that calculators being used daily in a calculus class and being immediately available on all classroom tests convinces students that the technology items are central to the course. "Will this be on the test?" is never in doubt in a calculator-based class, and that is one of the pitfalls in computer lab and computer demonstration-based instruction.

Writing requirements are one of the best moves in mathematics instruction. But we need to go the next step and require the students to produce formal, technical reports. That, more than any other goal, is the one that my engineering colleagues are asking for in the calculus classes on our campus. The newer graphic calculators are all easily linked with computer word processors, and it is an easy matter to insist that the students include graphs, data, and equations in their submitted reports.

Without a doubt, the computer station will always play a role in report preparation and large-scale investigations. It may be just as easy to enter large data sets into calculators as it is to enter them into computers, but the nature of the machines just lends itself to use of computer stations for large scale simulations, shared data bases, and formal demonstrations. The atmosphere of the activity is the over-riding consideration and not necessarily the capacity of the individual units. Yes, today's calculators have limited storage and processing capabilities, but wait till you see the next generation of units!

Personal Experiences

As noted earlier, I have almost a decade of personal experiences with the use of computing technology in mathematics instruction. Admittedly, that has been focused on graphics calculators, but many of the observations are equally applicable to any type of computing device. The instruction is not just *modified*, it is fundamentally changed! Those changes are significant and include:

- the approach to the topics,
- the nature of the class activity,
- the type of student experiences, and

• assessment procedures.

Computing changes the way that you look at topics. In my BC classes (before calculators), graphs were the end product. Hours were spent on amassing the information that produced the graph and few, if any, minutes were spent on discussions about the graph's implications. Now, problems start with the graph and *all* the discussion is about the dynamics in the represented relationship. Students now see functions as representatives of classes of processes that have underlying mathematical principles. Just as they see the *laws of nature* in their physical surroundings, they now see *mathematics principles* in their quantification models. The derivative is now "part of the package" and *not* a remote object that is treated apart from the underlying function. Rate of climb meters (vertical speed) and flight experiences give y' traces on the graphs a personal feel. The ability to zoom on graphs gives remarkable insights into local linearity and the graphical nature of differentiability. Initial results are treated with proper skepticism and the need for confirmation (rigor) is self-evident. The machines can easily deceive and lie to you, so skepticism is a constant companion.

I don't lecture anymore. Class is more of a discussion hour. Student questions dominate the coverage and very little time is ever spent on the mechanics of the calculator. The students learn from each other, and they seem to arrive to class early to share their experiences with each other. One "A" student accused me of *not* teaching mathematics. He came prepared to demonstrate his highly developed manipulative abilities, and he said in disgust that "I made them talk—just like in a history class!" and that I should return to his math class expectations and "have them sit there and write what the professor puts on the board."

The students are active participants in their learning. They leave the "drill and skill" mode and experience explorations. In my business calculus class we take a data-driven, technology-intensive, modeling approach. In a totally different course, every problem in the lecture notes starts with a set of referenced data. The student visually examines the numbers in a scatter plot, and then they are asked to reinforce their visual conclusion with a mathematical principle. If the data looks linear, then is there a situation where momentum is the over-riding force? Do constant differences confirm the linear claim? If the data looks parabolic, then is that same constant pressure driving the system? Do constant second differences support the conclusion? If it looks exponential, then why is it "feeding on itself," and do constant percent changes appear? If it looks logistic, then what is the nature of the carrying capacity, and why should it be "s-ing out?" If it seems to repeat in cycles, then what external force is there driving the oscillation? Having made these observations and reinforced them with a supporting principle, then technology gives them the coefficients for an analytic model. At that time, all the calculus techniques are available and ready for use against an equation that the student now personally owns.

Integration is equally different in an approach that is built on "accumulated change." Students in these "soft calculus" classes would have no problem recognizing the national deficit as the derivative of the national debt. Wouldn't it be nice if more voters were able to discuss the dynamics of economic policy and social issues with similar understanding!

Business Calculus is a vast waste land that is just beginning to experience the calculus reform movement. The new approach at Clemson University is only one of the exciting ways that technology allows this course to move into a modern era.

Tests are now only a part of the assessment procedures in our technology-based courses. Cooperative groups and team presentations are a major part of each student's grade. We felt intuitively that this was a correct approach, and then we knew it for sure when we heard a report that one Dartmouth College student said that *every* prospective employer in his job interviews asked "What meaningful team experiences did you have in your academic studies?" Group experiences are fundamental to our new business calculus course. We use the model of consulting firms making presentations to get jobs from firms that are "out sourcing" some of their tasks. The projects are described in a "request for proposal" format, and the groups become consulting firms bidding on the RFP. The presentations are made to a "Vice-President" of the firm, and a different VP hears the presentations on each of the different projects. In reality, the so-called Vice-President is a faculty colleague who is also teaching another section of the course, and we rotate serving as each other's Vice-President. The students see this as a reality exercise, and immediately find out that computer spreadsheet graphics are central to a quality presentation and a good grade. The projects and the presentations all call upon a wide variety of technologies.

The Future

Only a fool would try to forecast the nature of future technologies. It is frightening to realize that it was only two decades ago that *Visicalc* spreadsheets on an

64K APPLE II computer were new and exciting things. But if there is one observation to make that might have some hope of surviving, it would be the statement that **powerful hand-held computing units deliver technology to the masses**. Calculators and computers play key and different roles in my classes, and I have always been inclined to try all sorts of new things. But when I look back over my personal experiences, it was the personal/portable nature of the graphics calculator that changed the world for *all* of my students.

Things are happening so fast, and yesterday's dreams are today's products. We might be tempted to fall in the trap of saying that we have all that we need. Nothing could be further from the case. There are vast areas of mathematics instruction that technology has hardly touched. Geometry is the area that comes immediately to mind. This may not be the case for long when you see the power of geometric discovery in the new interactive geometry packages like *Sketchpad* and *Cabri*. The latter software is soon to be available in a hand-held device. Who knows what kinds of competing software and hand-held devices lie over the horizon? All I know is that the competition in technology products is very keen, and every day we see new products that overwhelm us with their power.

In computing products that enable us to do better what we have been doing, I see in our reluctant colleagues, Shaw's famous statement of "things that never were and asking *why*?" There is another whole range of products that we don't even think about, i.e., the other half of Shaw's statement about "dreaming things that never were and asking *why not*?"

Here's my dream, and I would love to hear yours. The business world is loaded with periodic data that managers must constantly live with. It comes at them as daily, weekly, monthly, quarterly, and annual reports. Analysis and forecasts control the firm's very survival. Wouldn't it be nice if we could feed the periodic data into a computing device and fit a regression model? Obviously, the underlying function would be some sort of truncated Fourier Series, but who cares as long as the computing device would give us statistical judgments on the quality of the predictability in the model, i.e., the expected values and the risks. I hope the calculator companies and software writers are listening.

Constructing a New Calculus Course

Deborah Hughes Hallett
HARVARD UNIVERSITY

When I first started teaching in this country as a foreign teaching assistant, I tried to figure out why the topics that were taught had been chosen. I persuaded several of my calculus students to take me back to their high schools, introduce me to their mathematics teachers, and let me sit in on some high school classes. From these excursions, I was left with the impression that there were some things that I might never understand. For a country of individualists that believed so little in central control, why were the syllabi so uniform? Why were there so few reasons available for the way the material was arranged?

Many years later at the "Calculus for a New Century" conference, I was exhilarated to learn that it might finally be possible to rethink our calculus offerings and design a course that made more sense for students and faculty, both in mathematics and in other fields.

After the Calculus Consortium based at Harvard was formed, we met to sketch out our vision of the new course. How could we make it lean and lively? Fewer topics were needed—but how should we decide which were essential? Going through a list of topics and choosing which to leave out is an enormously divisive process (one person's candidate for the axe always turns out to be another person's favorite), and implies that the original list has on it every topic that could possibly be important. Consequently we decided to start from scratch, taking no topics as given. To be admitted into the new course, a topic had to be proposed and agreed to by the whole group. To gain acceptance for a topic, the proposer had to be able to explain:

- Why a topic was clearly useful to students, either as a mathematical idea or in another field.
- Exactly what sort of problems students would be expected to be able to do on the topic.

The usefulness criterion was loosely applied, but forced us to discover and discuss the role of every piece of mathematics we taught in our students' academic lives. The problem criterion was a way of making sure that people didn't talk at cross-purposes: discussing a topic with possible homework and exam questions on the table avoids many misunderstandings.

It is important to recognize that "useful" does not have the same connotations as "relevant." We were not looking only for ideas that have real-world applications. We were looking for ideas which are mathematically important. We were looking for ideas for which we could make a good case that all students in calculus should spend time wrestling with. A good case might be made either on mathematical grounds or on grounds from another field. For example, the idea of a rate of change, of accumulation, and of approximating one function by a simpler one are, in our minds, central to a first calculus course.

Some topics changed their roles in the course. The Mean Value Theorem was often included in a traditional course in order to show that two antiderivatives of a function differ by a constant. Since the Mean Value Theorem was seldom proved, students were asked to take on faith a theorem they found difficult (the MVT) in order to prove a result they thought obvious (that two antiderivatives differ by a constant). Exam questions on the MVT were rare, and mostly concerned speeding offenses or computations of the MVT's "c." (The only question on the MVT in the "Calculus for a New Century's" exam collection asks students to calculate c for a quadratic.) We concluded that although many students may have learned the name of the theorem in the past, very few learned much more. Thus, we decided not to introduce the MVT at the traditional point in the course, but include it later when it comes up as a natural tool for estimating the size of an error.

To be able to have discussions about what topics are useful, one needs a good idea of what is needed in later mathematics courses (this is not hard to acquire) and a good idea of what is used in other fields. Since there are many other fields, answering the second question is not so easy. It is, however, essential. Most of the students we teach are not planning to go on in mathematics; most are taking our courses at the insistence of faculty in another discipline. If we cannot weave the ideas that other departments want taught into courses that are mathematically coherent as well as reasonable for the students at hand, we should not have accepted the responsibility for teaching these courses.

Several years before calculus reform started, the Harvard mathematics department re-designed Calculus II using significant input from other fields. The previous syllabus contained several topics whose purpose was not entirely clear. Thus we wanted to know what other fields needed. There are several methods of trying to find this out, some of which work better than others. One is to send faculty a questionnaire asking what topics they want. The answer usually comes loud and clear: everything (and by yesterday). Another is to organize a committee with members from both departments. Such committees can perform a valuable service—but it is ratifying and overseeing joint ventures, not in starting them. The initial contact is often best done in an informal way, where an individual mathematician meets with one or two people in another field—perhaps over lunch or coffee. Useful questions to ask are:

- How is mathematics used in the courses you teach?
- Where do students stumble mathematically in the courses you teach?

These questions usually produce a flurry of concrete examples (often on napkins) which are important guidelines for a new curriculum and a great source of problems. When the conversation turns to the topics in the course, the faculty from other departments usually want every conceivable topic included. Being accommodating leads to overstuffed courses. After establishing the (much too long) list of possible topics, a useful strategy is to propose a pair of topics and ask the other faculty: Which of these two topics would you want if there were only room for one? The answer you may get—a bit of both—is not an acceptable answer to this theoretical question, as the purpose is to establish an ordered list of topics. In practice, some cheerful persistence on your part will usually get the questions answered and the list constructed. Turning this collection of lists into a mathematically coherent course, which can be taught to real students, is a task that must be done by mathematics faculty.

When we asked these questions at Harvard, the response from faculty in different fields was surprisingly consistent. Their main complaint was, like the mathematics faculty's, that students couldn't use what they had been taught (making both sides wonder whether it had ever really been learned). As far as topics were concerned, the message was again consistent: most departments wanted more differential equations earlier (the exception was computer science). The fact that many of the other calculus reform projects chose to give differential equations a much more prominent place in first-year calculus underscores that what we heard was a widely held point of view—and one that mathematicians can ill afford to ignore.

As a department offering service courses, it is essential that mathematics departments hear and respond to such requests. It is not crucial—in fact it is not possible—to act on all of them. Some requests will conflict, others may lead to courses with no mathematical coherence, or courses that are unteachable. For example, we were once asked to put uniform convergence in calculus. Thus the final balancing of topics has to be done by the mathematics faculty. However, we must balance requests rather than simply dictate the topics we think important.

Calculus reform, besides providing the impetus to re-establish communication between mathematics and other fields, also made us re-think what students learn from calculus. The exams collected by the "Calculus for a New Century" conference make it clear that the calculus courses taught before calculus reform had essentially no theoretical component. The vast majority of the problems depend either on manipulative skill (the chain rule, integration by parts, etc.) or on having mastered a small number of template problems (volumes by discs, related rates, etc.). Success on such exams was often achieved without even a basic understanding of the concepts—and this lack of understanding became apparent when students moved into subsequent courses. Traditional texts did, however, contain some theory, though it was ignored by virtually all the students as it was practically never tested. Thus, the reformed calculus courses which asked students to explain their reasoning are described by students as "harder" and "more theoretical" than the courses they have taken in the past. At the same time, some mathematics faculty are saying that the reformed courses are easier. In fact, reform courses generally do involve significantly more thinking on the part of the students than the traditional courses. Neither traditional nor reformed calculus courses are theoretical in the usual sense of the word. However, if we judge a course by the problems done rather than by the theorems displayed, then the reform movement has taken a small step in the direction of putting some mathematics back into calculus courses.

Thinking about Learning, Learning about Thinking

David A. Smith
DUKE UNIVERSITY

This is a personal essay—not a research paper—about how thinking about learning affects the way we teach. For the most part, I am writing about my own thoughts and behaviors. When I use "we" usually I am referring as well to my colleagues in Project CALC at Duke and elsewhere, especially my co-director and co-developer, Lang Moore. Occasionally "we" will mean all who teach calculus, and the context should make this clear. All opinions expressed herein are my own, but the words in which they are expressed in some cases were first uttered by others. I apologize in advance for failing to include proper attributions. Those of us in the calculus reform movement have shared so many ideas with each other that no one knows who first said what any more.

Participating. For more than 30 years, I have announced to each new class that mathematics is a participation sport, not a spectator sport. Before the mid-1980s, what I meant was that students were supposed to take my carefully chosen words of wisdom back to their dorm rooms and work hard on an assignment in order to understand the mathematics *du jour*. Only a few did that, of course. Most did not consider it important to participate—other than on routine homework exercises—until it was time to "study" (i.e., cram) for a test. And then their objective was not *understanding* but a grade.

My past behavior was analogous to a coach using the practice session to diagram plays on a blackboard—so that players would not have to read the playbook—and then sending them away to practice alone until game time. When it was time for the contest against the other team (Ignorance), they were expected not to communicate with each other, but to run their plays in isolation.

For most of my career, I had little sense of mathematics as a *team* sport, so perhaps my analogy for past classroom activity should be with tennis or golf instead of basketball or soccer. My practice time was devoted to lectures on the finer points of backhand or approach shots, with instructions to practice alone—which students would mostly neglect to do—until just before the tournament. Indeed, mathematics-as-participation-sport is both an individual and a team effort, but it doesn't make much sense to practice either type of activity alone. It makes even less sense to devote all the available coaching time to making students listen and take notes.

The metaphor of coaching is common among those thinking about and working on reform. We know what it takes: a mixture of short bursts of instruction and/or demonstration, longer practice sessions, motivational pep talks, the challenge of real competition, and assistance with in-game strategy. But we can't coach players who won't come out for the team, or who sulk instead of play, or who quit when they don't make the starting team. To keep our students engaged in the game we coach, we have to convince them that the pain and struggle are worth it—that the game can be fun to play, that it is more fun if you work harder at it.

Reading. When we embarked on Project CALC, we weren't at all sure that a reformed calculus course should have a textbook. Weren't those thousand-page monsters more of a problem than a solution? For a generation or more, everyone teaching calculus collaborated in demanding (or writing) textbooks that would help our students *pretend* to learn mathematics by memorizing rules for moving symbols in response to set types of problems—"template exercises"—that have no relationship to mathematics as practiced by anyone else. We had an unspoken contract with the students that *we would pretend to teach if they would pretend to learn*. The primary aid we provided for carrying out their end of the bargain was a book in which the words did not have to be read. It was sufficient to match each exercise to a "fully worked out example" that would reveal the sequence of symbol manipulations sufficient for getting the answer in the back of the book.

As the Project CALC development progressed—and as we observed students learning to solve substantial problems—we found ourselves writing more and more background information for which there was not enough time in the classroom. As we wrote, we rediscovered the proper role of the textbook as a source of information. But to serve that role, the book has to be read. In mathematics, it has to be read with pencil and paper at hand—and now a graphing tool as well—responding to frequent challenges. It is not enough to read through pages of carefully reasoned material, dutifully agreeing with the author, and then expect to be able to do as the author did. Given our students' general lack of intrinsic interest in the course, we had to present new information in engaging contexts. Thus was born *The Calculus Reader*.

Alas, it was not to be. Word came back from the publisher's representatives trying to interest faculty in preliminary editions of the *Reader* that, if the book signaled "read" on the cover, faculty would not even look inside, let alone consider adoption. So our labor of love will appear in first edition (before this paper appears) under a new title that will be less scary to our colleagues.

That still leaves the problem of how we get students to read about mathematics when they never had to do that before. The problem is not unlike getting them to write about mathematics when they never had to do that before. Many others have addressed the latter issue with various approaches—some quite successful—to Writing Across the Curriculum. I have written elsewhere of having to invent Reading Across the Curriculum—possibly a harder task, because no one wants to admit that otherwise qualified college students can't read.

Well, in fact, they can read—and write—but someone has to teach them how. Lacking freshman reading courses analogous to our ubiquitous writing courses, we have to do it ourselves. Our first step has been to do reading exercises in the classroom. For example, early in the course we might assign small groups to read two pages of text to each other, working the exercises as they go, with an expectation of being called on to report before the period is over. Under supervision—and with an appropriate reward or penalty—students find they *can* read a well-written book. Once they *believe* it is possible, it *is* possible. They continue to need incentives, such as pop quizzes focused directly on reading assignments. If they continue to use their new-found ability throughout the course, then our textbooks can serve their intended purpose—the same purpose books have served since Gutenberg.

The payoff for teaching students to read is that we can stop worrying about *covering* the syllabus—properly the students' job—and concentrate instead on *uncovering* it.

Writing. How do we know what other people think? In the academic world, we read what they write, we listen to them speak, and we ask questions. Verbal interactions are often informative, but they are not necessarily a "clean" look at the thoughts of others because our own thoughts are part of the process. Usually we don't care. However, if we really want to know whether another mathematician has proved a theorem, say, we want to see the proof in written form with all the details—and no intervention by others, including ourselves. Given that you have read this far, you may actually care about my thoughts on learning, and you are getting as good a look at those thoughts as I can provide.

Why should it be any different with students? If we are to take seriously our goal of having them understand concepts, we need a window on their minds. They have to tell us what and how they are thinking. We can listen to them talk, and we can ask them questions. That's a start, but our own thoughts are inevitably part of that process. The cleanest window we have is student writing.

Faced with a demand that we see their thoughts, students will find—consciously or unconsciously—many ways to darken or obscure the window. Some of those tactics have actually been taught to them, e.g., not showing up in their expositions in the first person. Some tactics are intuited, such as using passive voice or expressing their actions in nouns rather than verbs—so they don't have to reveal themselves as the agents of any action. Other tactics include using apparent nouns as pronouns without clear antecedents—e.g., "the function" when there are two or three that might be the antecedent—or using non-specific verbs, such as "manipulate," which can refer to any mathematical technique they have ever learned.

Even in evasive mode, students often reveal more in their writing than they think. For example, a favorite expression is "We received the following answer: ... ". First, this reveals a belief that the only important thing is the answer, not how they found it—but they have to satisfy the assignment by putting in some words. Second, their choice of verb reveals that they believe mathematics is not something they do but something that gets done to them.

We have to make it clear to students—early and often—that we will not tolerate evasion, obfuscation, or misdirection in their writing. They must actually reveal their thoughts, and they must do so as clearly and directly as possible. However, in order for them to trust us enough to do this, we must respond in supportive—not punitive—ways. We must convince them it is in their interest to reveal their ignorance and misinformation—along with their valid knowledge—so that we can help them replace ignorance and misinformation with valid knowledge.

This is hard for both faculty and students because both have long seen each other as adversaries in the process of assessment—it's us against them on both sides. If the only things we ever mark on papers are mistakes, where will they get the idea that we are on their side? If we really believe that only a few can succeed at the highest level—that most students are not very good (implicitly, can't become very good)—how will students come to believe in their potential to succeed?

Most schools have people whose job it is to teach writing—and a requirement that students learn from those folks. Writing programs that take seriously the concept of writing to learn can be very helpful to faculty who need to learn techniques for positive responses that will induce students to write clearly about their thought processes. Of course, this means we have to initiate a contact across what looks like a very high wall. But once that contact is made, the folks on the other side of the wall are usually delighted to hear from us. Often it turns out they never believed someone on the science side of the campus would actually care about them. If they're any good at all, they can make the task of reading and responding to student writing much less burdensome—often even fun. It's time-consuming work at first, but with practice—like anything else—it becomes quite doable.

Calculating. How much calculation is necessary in a calculus course? After all, that's what the word means: a system of calculation. We want our students to have good "symbol sense"—but do we know what that means?

This was not much of an issue with the traditional course because the whole course was about calculation. We assumed—with no theoretical or empirical justification—that students could learn calculus by watching us calculate, and that we could learn what they knew by watching them calculate. Our job was to find the best examples to show them and then to test them with examples that would surely show whether they "understood" or not.

Thinking about who our students are and how they relate to mathematics, we have found it much more productive to embed the need for calculations in contexts that students find meaningful. When we start with a problem students might actually care about and find that we need a new technique to solve the problem, there is a motivation for learning that new technique. "You're going to need this later" never was much motivation, especially if it turned out not to be true.

On the other hand, there is some justification for a modest amount of unmotivated and uncontextualized computation—if it's not too difficult—because students whose entire mathematical experience has been symbol-pushing find a comfort level in doing familiar activities.

Apart from making students feel comfortable, we think that some amount of symbol manipulation is necessary in order to understand what machines do for us and to monitor the reasonableness of machine-generated answers. We are convinced that will continue to be true even in environments where essentially all calculations are done by machine. Lacking any theoretical base for deciding how much is enough, we continue to address this problem empirically. However, we have found through our use of gateway tests that most students can learn appropriate symbolic skills on their own and from each other, with relatively little class time or text space devoted to this.

Assessing. We know that tests can measure only a small portion of what we really want students to know. Furthermore, the usual environment for testing is very artificial relative to the environments in which they will later have to demonstrate their knowledge. If we are to teach, say, how to solve realistic problems, then we have to assess students on their ability to solve such problems—not just on the small pieces we can ask about on tests. If we are to teach conceptual understanding, then we have to find ways to measure such understanding. And if we really want steady, sustained effort from students, it will not do to assess only every four or six weeks.

Fortunately, the solution to all these problems is the same: Make every learning activity an assessment activity as well. For this to mean anything, there has to be a lot more going on in the classroom than just the instructor talking. There is no way to assess listening, and it makes no sense to assess note-taking. On the other hand, when students are actively engaged in learning, everything they do can be assessed, formally or informally. For example, an instructor interacting with a problem-solving group is both facilitating the learning process (coaching) and observing how well they are doing (keeping score). This is likely to be an informal assessment—it may not be assigned a visible grade, but it certainly informs the instructor's subsequent grading process. The outcome from the problem-solving process—a blackboard presentation or a written report, say—can and should be the object of a more visible assessment. Students need feedback on everything they do, the more frequently the better.

Let me emphasize that assessing and grading are two different things. I have already noted that informal assessment processes are likely not to be assigned grades. But even formal processes can be for "practice." This is especially important early in a term when students may have to learn such things as reading, writing, and working with a group. If they know that the same activities will later be assessed in the same ways—and that the resulting grades will "count"—they will take the practice activities just as seriously as the "real" ones.

Knowing. For as long as I can remember, the dominant philosophy in college level mathematics education has been what I call *transmissionism.* Teachers know and students don't. Teachers are knowers who transmit knowledge to learners. Then learners know as well. But never quite *as well*, because, after all, we're the experts and they're the novices. So what we see when we test the effectiveness of our crystal clear lectures is a pale shadow of the brilliance we emitted. And that's their fault, not ours.

As soon as we insisted that students write about their thought processes, we realized that transmissionism is an indefensible philosophy. In fact, what our students *know* about mathematics, for the most part, was not transmitted to them by any teacher or textbook—none of us would want to take credit for it. And yet, that knowledge was getting them through courses with passing grades, sometimes with excellent grades. We call this form of mathematical knowledge *coping skills.*

Where do coping skills come from, and how do they work? Early on, we were discovering a constructive proof of the dominant paradigm in education, which we now know is called *constructivism.* Learners construct their own knowledge in response to challenges to their current state of knowledge. Given the steady stimulus over many years of having to pass tests on dimly understood material, students continually discover and share ways to get apparently correct answers—without wasting time on understanding. Once we understood—even dimly—how this works, we were able to focus on challenges that demanded understanding as an integral part of finding sensible answers.

Differentiating. One of Sheila Tobias's well-known books is *They're Not Dumb, They're Different: Stalking the Second Tier.* "They" are otherwise successful students, graduate students in non-science disciplines who had not been successful in college-level science courses. They were asked to enroll in such courses and keep journals of their experiences, documenting what worked and what didn't—mostly the latter. In a nutshell, entry-level science courses for the masses were being taught as a trial by fire by faculty who were among the few to survive such experiences when they themselves were students. The graduate students were able to identify the factors that interfered with their success in understanding science. Much of the book is about documented techniques that accommodate a wide variety of learning styles and lead to broad-based and substantial understanding, both for those who will and those who will not become scientists. Furthermore, these techniques open doors to science careers for traditionally underrepresented groups of students.

What I learned from Tobias is that almost all the students in our calculus courses are in a mathematical second tier, even the science and engineering majors. At Duke the first-tier mathematics students place into Calculus III or beyond, and on most campuses they are a tiny percentage of the calculus enrollments. The message about the overwhelming majority of our students is *they're not like us*—and that doesn't mean they're dumb. If we use ourselves as models for how students learn mathematics, we always get it wrong.

From the outset of Project CALC, we knew from our own and others' experiences that there was hope for accommodating a broader variety of learning styles by using small group activities, laboratory experiences, and writing. But we had no idea how any of this might be related to gender differences. Furthermore, we were not aware that we had a gender-discrimination problem in our traditional course, so that was not a problem we set out to solve. When we first saw women succeeding in ways we had not seen before, we attributed this—on average, not for every individual—to the greater verbal content of our course and to ways to succeed by expressing ideas in writing. But we also observed women speaking up more in small groups than we had seen in whole-class discussions. Later in the course we were seeing more participation by women in the whole-class discussions as well. And all of our teaching experience suggested that—on average—those who spoke up in the classroom were most likely to succeed in the course.

Later we learned of research pointing to a potential danger for young women of placing them in groups in which they were a minority—e.g., a single woman in a four-person group. I adopted a policy of not doing that until I could see enough evidence of the strength necessary to hold her own and not be suppressed by male peers. I often see this kind of growth in less than a year, sometimes in less than a semester.

Every time I mentioned these gender-related observations in a faculty workshop, a female participant would respond,—"You know, you really should read *Women's Ways of Knowing*." The third time this happened, I added this book (by M.F. Belenky, *et al.*) to the Project CALC collection and read it carefully. I learned that the phenomenon of first speaking up in a small group and then in a larger one is called "finding her voice." More importantly, I learned much more about stages of development and how they differ by gender. In particular, I gained a better understanding of the work of William Perry on learning stages—all based on studies of male students at Harvard.

Learning about learning stages has helped us see that some of the things we attempted early on were inappropriate, and we changed them. It also helped us understand better why so much of the traditional course and its methods are inappropriate for the intended audience—at least if we really want all those students to succeed. That course was an excellent filter precisely because it was so mismatched with such a large segment of its audience.

Learning. Learning stages are, by definition, not permanent. Indeed, it is not unreasonable to set a conscious goal of trying to get students to make the transition through one or more Perry or Belenky stages. We have not stated our goals that way, partly because such goals would not be recognized by most of our colleagues, but mainly because the stages don't describe adequately the specific mathematical and other abilities we intend for our students to develop.

Learning *styles* are also not permanent, although many teachers and students act as if they were. In fact, it is important for a learner to develop a repertoire of learning styles, and it is important for teachers to encourage that development. I occasionally have a student, usually male, who complains (not necessarily in these words) that our course does not accommodate his preferred learning style, which includes working alone and being rewarded or penalized solely on his own ability to "solve problems," not to write about the process. Such a student is seldom swayed by the argument that he will have to make a living by working with other people and by explaining his work to others—that it is time to learn how to do those things. I can empathize, because I retained the working-alone mode for much of my professional career—never letting others see me make mistakes if I could avoid it. My professional life became much richer when I learned that I could share work and learning (and mistakes) with other people, and I can quite honestly recommend that my student not wait as long as I did to make this discovery. I see no disservice in requiring that student to work with others and to explain what he is doing.

Actually, this student and most of his peers have a lot of *unlearning* to do before any significant learning can take place. In particular, we have to make the price too high for them to continue their mode of constructing coping skills for passing the next test. There are two things we can do about that: (1) stop making test scores the dominant part of their course grades, and (2) stop giving tests that can be passed by coping skills in place of understanding.

One way we encourage students to learn new learning styles is by placing them in heterogeneous groups to solve problems that may be too hard for any individual in the group. Each student brings to the group activity her or his prior problem solving experiences, and the mix is often richer than the sum of the parts. As students learn from each other, they also learn the value of others' learning styles, and they begin to add aspects of those styles to their own view of learning.

Modeling. As already noted, when we use ourselves as model learners, we always get it wrong. That is, if we reflect on how we learned mathematics and design our courses accordingly, those courses will be appropriate only for mathematics students who think and work like we did.

However, there is a sense in which we *can* use ourselves as model learners—but only if we focus on subjects in which we never became experts. I can comfortably compare the mathematical successes of many of my students with my own experiences in foreign languages, athletics, and music. There is no danger that anyone will consider me a linguist or an athlete, and I am at best an *amateur* musician. My native talents in these areas range from none to modest, but I have managed some small successes that have been personally satisfying.

Of these three areas, language is the only one in which I have any academic experience. My satisfaction in that area was almost entirely in the form of consistently good grades. I suspect I acquired those grades by twisting the subject into one I understood—mathematics—and focusing on structural analysis, patterns, and memorization. Much of what I memorized is now gone, but what I learned of structure and patterns in Latin, French, and German was quite beneficial for understanding English. On the other hand, I failed utterly to achieve one of the primary goals of language instruction: ability to communicate. Furthermore, whenever I entertain the rash idea of uttering a sentence to a native speaker of another language, I panic.

Here is one model for the experience of many mathematics students that we may consider "pretty good:" able to approach the subject in some way on their own terms (often terms not revealed to us), able to get good grades (perhaps by memorization and heroic efforts at test time), perhaps able to use some part of the subject in another area, but essentially unable to achieve our goals (or their own) for "understanding," and likely terrified at the prospect of having to carry on an intelligent conversation about mathematics.

A second model I draw from my athletic experience. I have little natural athletic ability, but I had good coaching from my father and occasionally from others. I was often the last picked when sides were chosen for a pickup game (any sport), but my father always made sure I got reasonable playing time on teams he coached, and he spent many hours with me on the golf course. There were no special benefits from being related to the coach, other than being in the right place at the right time. In a long and productive coaching career—with no monetary compensation—he touched the lives of perhaps two thousand young people, and they were all "his kids." A few are now professional athletes, but that was not his goal.

Here's the model for student learning: A loving and dedicated coach was able to get me to challenge my deficiencies and achieve beyond my innate abilities—not by just telling me what to do, but by sticking with me through many frustrations and by convincing me that it was worth the effort to try and to keep trying. In fact, my modest accomplishments with a wide range of individual and team sports have been much more fulfilling than my good grades in language courses. This model has also been much more relevant to my relationships with students in the last decade or so. I'm a slow learner, but eventually I realized that I had a much better career role model in my father than I had in my math professors, most of whom were excellent. With me as student, his job was much harder than theirs.

I get a rather different message from my experience with music. This is an area in which I have a little talent, although certainly not enough to make a living. I also have very little formal training and certainly would have benefited from more. But when I started lessons with a serious musician, neither he nor I could stand each other very long, and I lacked the discipline to practice as seriously as he demanded. The opportunity never came around again. On the other hand, for more than 40 years I have been in instrumental and choral groups directed by true musicians who had a talent for drawing out better performances from amateurs than we could accomplish on our own.

I see in this relationship with music a similarity to what I see with many students who don't think of themselves as mathematicians-in-training, but who are quite capable of individual or group problem solving, as well as conceptual understanding, when properly coached and conducted.

One might argue—correctly, I think—that learning mathematics is not like learning in any of the three areas just mentioned. My point is not that we should model mathematics education on language education, but rather that we can use our own experiences ranging from utter failure to satisfaction—not our areas of professional success—to get some sense of how most of our students relate to the discipline we are trying to teach.

Experimenting. There is a well-known theorem about educational experiments called the *Hawthorne effect*: Experiments always succeed. This observation derives from studies that show people work harder and more effectively if someone is giving them attention that seems to be "special." And of course that is always the case when they are the subjects of an experimental approach.

Critics of calculus reform raise questions about whether students wouldn't learn just as much from traditional courses and traditional texts if they spent as much time on them and worked as hard as they do when they are in experimental courses. The questions are silly, for two reasons: (1) Reformed approaches have been in use for five or more years at some schools, and many of these schools no longer have traditional courses. In those cases, reformed calculus is no longer experimental. (2) We already know that students won't spend as much time or work as hard on dull and meaningless courses—that's one of the primary reasons for reform.

Furthermore, discussions of time and effort tend to focus on student-reported averages without any discrimination among different types of students. My recollection of traditional courses is that the students who spent the most time on the course were those who struggled and never "got it." At the other end of the scale, many of the A students were breezing through a course in which they had already succeeded once, so they spent little if any time on it outside of class. In our reformed course, time and effort are much more correlated with success, and labels such as "strong" and "weak" much less so.

There is, however, a warning for reformers and would-be reformers in observation (1). It is possible, once the new text and classroom practices become routine, to slip back into a rut, always doing the same things in the same ways. Faculty have an uncanny knack—and lots of practice—at making interesting material dull. Fortunately, the collective reform movement has found many different ways to make the study of calculus fresh and interesting, and the materials from any one project are likely to offer much more variety than can be used in any one year. It's important to keep varying the course to keep from getting stale.

Put another way, the Hawthorne Theorem is not an indictment of reform, it's a prescription for success. Its corollary is that one should experiment with something every time a course is offered. Faculty who are always trying something a little bit new are also always giving their students "special" attention to see how the new thing works. Students will respond to that attention with more work and more time on task. In a reformed environment, this leads to more students achieving at a higher level.

Summing Up. I have grouped my thoughts into ten sections, each labeled by an active verb. Here I offer a sentence or two of summary for each verb.

Participating: We must keep students actively involved in their own education.

Reading: If students don't know how already, we must teach them how. Otherwise *they* won't be able to cover the syllabus.

Writing: Students must write to reveal their thinking. Our colleagues in writing programs can help with this, but we must reinforce their efforts.

Calculating: Even if machines can do it all, some calculating is necessary to develop symbol sense—and it makes students feel good.

Assessing: Every learning activity is also an opportunity for assessment. Students need frequent feedback—formal or informal—and teachers need frequent updates on their understanding of what their students have learned.

Knowing: Students construct their own knowledge. Our job is to make sure they do it right.

Differentiating: We must provide routes to success for students who are not like ourselves.

Learning: We must pay attention to learning stages and learning styles—and realize that neither is permanent.

Modeling: To understand how our students relate to mathematics, it is useful to recall our own efforts in areas where we never became experts.

Experimenting: All experiments succeed. Therefore we should always experiment.

Part II:

Planning

It's easy to say you want to throw a great party. You might think it's as easy as choosing the flavor of pie and ice cream. But what kind of pie will your guests want? What flavor would be the best for ice cream? Who will come to the party? Will they all enjoy the same dessert? You could make the pie and the ice cream from scratch, with fresh ingredients and use a new recipe. Maybe it would be safer and easier to buy something from the store and the bakery. Should you send invitations and describe the party carefully so people know what to expect? Do you know what to expect? Maybe you should hold a smaller party first to experiment. Why not invite just your close friends who will not be offended if things don't taste right? And what about the music and dancing? And what about people who want to play board games or talk with others quietly rather than have to be involved in group activities?

This section is not about giving parties, but about planning to make changes in your department's calculus program. So why begin with a paragraph on a frivolity like parties? The truth is that at many institutions faculty members will spend less time on planning changes that reform the calculus program than the time professionals would spend on planning a large party with five-hundred invited guests. Why is that? In part, the reason for the failure to devote adequate thought and time to planning for change is simple. When people start to change they do not always consider that the changes involve more than a reordering of the current topics, the introduction of some new topics, and the application of technology with these changes. Certainly these are a part of the changes that will come with planning. However, to make changes in a calculus program that will last longer than the usual text adoption and are more than superficial, a faculty needs to consider, with some care, a range of questions and issues.

Whether short or long in final form, most effective plans for change recognize the reasons and goals for making change as well as the related objectives and reforms. The party analogy may not be a perfect model for planning, but it can help those just starting to consider change to think about making choices that respond to their students' needs and reflect their own capabilities.

How does a department start to plan? Who does the planning? How do you form a plan? When do you have a plan? How do you implement a plan? How and when do you revise and change a plan? None of these are easy questions, and there are even more questions about planning that haven't been mentioned. For each department the development of a plan for change will depend primarily on the talent and energy of the department members who support change, and the general willingness of the community to consider and participate in the redesign of their program. "Constructive ownership" is perhaps one of the more over-worked concepts today, but it does seem true that when people participate actively in the process of planning change, the overall results in product and enthusiasm for implementation are more positive and durable.

The remainder of Part II should help you understand and proceed with your own planning. It begins with a collection of comments reporting different experiences in trying to bring change to calculus programs. These come from a small sample of faculty representing some of the variety from across the nation. The comments are presented in the format of a panel sharing experiences in changing the calculus program at their institutions. This is followed by a separate article by Morton Brown describing the planning effort at the University of Michigan, Ann Arbor. The Michigan project is one of the largest planned restructurings of a calculus program, and was funded in part by the National Science Foundation.

Part II closes with an essay on common features and stages of planning for change. This gives an overview of some stages involved in planning change. At the end of the essay you will find a checklist of items to watch for that may help you avoid some painful bumps and bruises while developing and implementing your own plan for change.

Planning and Change: Experience Reports

Martin Flashman
HUMBOLDT STATE UNIVERSITY

We posed a number of questions to faculty at institutions that have undertaken department-wide changes in their calculus programs, asking them about the planning process and the early stages of making changes at their schools. Here, organized much as they might be in a panel discussion, is a selection from the comments we received. Each response has been identified with the school of the respondent as follows:

Cal Poly will indicate the remarks of Claudia Pinter-Lucke from California State Polytechnic University, Pomona, California.

Rio Hondo designates the comments of Carole Fritz, Rio Hondo College, Whittier, California.

Anon marks the comments of a contributor who requested personal and institutional anonymity.

Mount Holyoke refers to comments by Harriet Pollatsek, Mount Holyoke College, South Hadley, Massachusetts.

Ole Miss denotes the responses of Charles Alexander who teaches at the University of Mississippi.

Oakland indicates the remarks of J. Curtis Chipman of Oakland University, Rochester, Michigan. These were taken from his preliminary report, *A Process for Departmental Change*, January, 1994, which in turn was based on a note distributed in the MER Calculus Reform Resource Collection, *Educational Reform at the Department Level*, in 1992.

Background

Describe your school and the way calculus was taught there before the change process started.

Cal Poly: California State Polytechnic University, Pomona, is a large public university east of Los Angeles which serves 19,000 students, with 2000 residential students. The largest colleges on campus are Engineering and Business. As a result, the Mathematics Department is largely a service department. Six years ago, there were three calculus tracks (there are four now). The track which serves the most students is the five-quarter calculus sequence for physical science and engineering majors. It consists of three quarters of single variable calculus and two of multivariable calculus.

In 1989, to begin the sequence, a student needed a C or better in both college algebra and trigonometry, or an appropriate score on the mathematics diagnostic test. This has since been expanded to a B or better in college algebra and trigonometry, or a C or better in pre-calculus, or an appropriate score on the diagnostic test. During any given quarter there are four to ten sections of each quarter of engineering calculus. In addition, similar criteria for placing students were used in separate calculus courses for engineering technology majors, business majors, and biology majors. Six years ago, all sections of all calculus courses were taught from traditional textbooks (all sections of a particular course using the same text) with the traditional lecture method. Technology was absent from the classroom.

Rio Hondo: From 1966 when Rio Hondo College first opened its doors as a community college in Whittier, California, until August 1993, calculus was taught in the "traditional" way. This mode consisted of a teacher standing in front of the class lecturing, with occasional use of the blackboard. The text had to weigh at least 25 pounds, and be written by someone named Stein or Anton or Larson. A "radical" instructor might use an overhead projector, or let his students use a scientific calculator.

Anon: We are a four-year undergraduate institution with quite selective enrollment. We have small sections of our classes (typically 16–21 students) and chalkboards around the full circumference of our rooms. We use only full-time faculty. We have a history of pushing the use of technology (scientific calculators required in fall of 1975, programmables required in 1982, and PCs in 1987). All of our students for the past two years have had ready access to *Mathematica*. Before that we used *MicroCalc* in beginning courses and *MathCad* in others. *Derive*, *Matlab*, and others have also been used. With the exception of our attention to technology, our calculus courses were very "traditional." All of our students are required to take two semesters of calculus regardless of their major, so our courses have something of a "liberal arts" flavor even though engineering majors take the same first-year calculus.

Mount Holyoke: Mount Holyoke College is a small (1900 students) liberal arts college for women, with a strong tradition in mathematics and science. We are a department of 10.5, including mathematics, statistics, and computer science. Calculus has always been taught in sections, rather than large lectures, and faculty typically spend a lot of time with students in office hours in addition to class time. Often, the department has chosen a common calculus text, but not always, and instructors have always had considerable classroom autonomy, even in introductory calculus. In 25 years at Mount Holyoke, I can't ever remember using a particular calculus text for more than three or four years.

Ole Miss: The University of Mississippi is the oldest comprehensive university in Mississippi with current enrollment on the Oxford campus of approximately 10,500 students. The Department of Mathematics has a budgeted faculty of 16 full-time professorial rank faculty and offers the B.A., B.S., M.A., M.S., and Ph.D. degrees, and has 70 undergraduate and 28 graduate mathematics majors currently enrolled.

Approximately 300 students enroll in the calculus courses each year. All courses at the level of calculus and above are taught by faculty with the doctorate, or, on rare occasions, by a doctoral student near completion. Before introduction of calculus reform, the Department's calculus program followed a traditional syllabus covering most of the material in Finney and Thomas's *Calculus* in a four-semester sequence, using the lecture method and no technology.

Oakland: Oakland University is a publicly-funded institution with undergraduate and graduate (headcount) enrollments of approximately 11,000 and 2,000 respectively. The residential student population is quite small, about 1,500. The Department of Mathematical Sciences accounts for approximately 10 percent of the total credit delivery; there is a full-time faculty of 27, with an additional mixture of (masters level) graduate assistants and part-time faculty which brings it to a full-time equivalent staff size of 37. The department's culture is very democratic; faculty expect to be consulted and to give their approval on all major issues. As a result, there is a departmental faculty handbook which runs to just over 70 pages! While research is the top priority, instructional responsibilities are taken very seriously. Almost all introductory classes are conducted in a multiple section mode where class sizes typically range from 40 to 80 students. Precalculus is the only exception to this and uses a large lecture-recitation format. Only full-time faculty teach the traditional calculus sequence. Multiple section courses have always been taught from a common syllabus with a common text, with coordination amongst the instructors to provide for consistency amongst the various sections. Common final examinations are typical. Amazingly enough, these courses, taught by committees whose memberships changed frequently, often looked like courses taught by committees. Policies, procedures, and success rates tended to vary considerably from semester to semester. At the same time that this was going on, we were also experiencing the high attrition rates common to many public institutions whose admissions policies are closer to "open" than "selective." (In our traditional calculus, 55 percent of the initial class receiving a 2.0 or better is a "good" year, whereas a number like 40 percent, while disappointing, is not a disaster, for we have seen worse.) Dissatisfaction with this state of affairs was increasing within the department. It's not a lot of fun to teach in an emergency room, and a number of us could not help but notice that some yellow lights were also starting to flash both in the client disciplines and our own dean's office. All of these factors encouraged us to take the plunge and enter the thicket of delineating the boundaries between the teaching faculty's control and the department's control in our 100–200 level introductory courses.

Initial Phase: How Did Change Start?

Cal Poly: Change was prompted by three events:

1. The introduction of the successful Academic Excellence Workshops program.

2. Mini-conferences sponsored by Mathematics Departments at other California State Colleges about technology in mathematics instruction.

3. The creation of a new pre-calculus course at Cal Poly Pomona (none existed at that time). By 1991, a few instructors used calculators in trigonometry and linear algebra sections or brought the computer into the classroom for demonstrations, one instructor had a weekly cooperative learning section attached to her calculus course, another did not lecture but used Learning Through Discussion, and two instructors included long-term projects in their assignments. Course notes were written and adopted for the new pre-calculus course in which calculators and group work played an integral role. All these activities were performed as individual endeavors. They were not greeted with hostile or friendly reactions because there was no discussion in the halls or at department meetings about these efforts. There was no release time for development, and most of these instructors did not share their results at professional meetings.

Rio Hondo: In the fall of 1991, one of our faculty members, Bev Abila, attended a session on calculus reform at the AMATYC conference in Seattle. Then in the summer of 1992, she attended a week-long workshop on calculus reform. At about the same time, Paul Moreland went to a workshop on graphic calculators and got real excited about the visualization and effectiveness of these machines. He even experimented with the pre-calculus course and used graphing calculators extensively in his class.

During this time period, I was teaching second-semester calculus and really agonizing over it. It was extremely painful for me to spend two-thirds of a semester on integration techniques and series and sequences. I really thought there had to be a better way to do this. The students were successful at playing with the algebra and notation, but they didn't seem to have much understanding of why we were doing these things or where we would ever use them. In all the calculus classes, I felt that we were covering an enormous amount of material in a very superficial way.

I don't think "outside" people can come in and tell math faculty what they should do; they automatically get their backs up. Math folks like to think that any good ideas originated with them. You need two or three assertive, influential members of the department to initiate the change.

Anon: In 1990, our department head and two pedagogically-oriented senior faculty members began considering serious reform of our calculus courses. This *ad-hoc* committee of three considered a wide variety of options. They recognized the need to establish clear goals for guidelines. This is what they came up with: Change the thrust of our calculus course from one of basic technical skill mastery to one of problem formulation and solving using the tools of calculus and the new software technologies. The department head determined that a large scale experiment would begin in the fall of 1991. The committee began to determine whether the reform should be based on improved teacher training, modified classroom environment, new text, or some combination of these. It was decided that a change in text would be necessary to send the message that we were serious about change. Also, for the experiment, only teachers positive about the changes would be used. The initial idea was for this course to become the main one in 1992. We also recognized the need for a significant modification to the way we looked at exams if we were to assess success in making progress to our new goals.

Mount Holyoke: Well before the calculus reform movement began, we were making occasional use of computer laboratory exercises in calculus. In 1981, my colleagues George Cobb and Bob Weaver created a January mini-term course in which students learned to write simple BASIC programs to illustrate ideas from calculus and do more realistic applications. It was taught again in 1982 and 1983, to about 15 students each time. In 1983, Bob Weaver wrote a calculus utility for the Apple II (Compucalc), and several of us used it in first- and second-semester calculus classes. I think it would be accurate to describe these uses of technology as additions to the traditional course, rather than as fundamental changes in how we were thinking about the teaching of calculus.

In 1986, we invited colleagues from Hampshire College to run a weekly seminar for our faculty on their innovative course for life and social science concentrators (and curious humanists) that emphasized modeling and numerical solution of differential equations. At the same time, we read Peter Lax's lovely essay "On the Teaching of Calculus." Here "we" refers to our entire department; all of us attended the seminar on the teaching of calculus and were quite interested in exploring ways to change our teaching in more fundamental ways.

Ole Miss: A senior faculty member spent the fall of 1989 on sabbatical in order to meet with prominent developers and implementors of mathematics software for instructional purposes. He also attended several conferences devoted to use of instructional technology in mathematics, including the annual Conference on Technology in Collegiate Mathematics, and the St. Olaf Computer Algebra System Workshop. He has attended short workshops on *Maple*, *Derive*, and Writing-to-Learn-Mathematics, and received a Writing-to-Learn-Mathematics summer grant from the College of Liberal Arts in 1990. During the Fall of 1990, he taught one section of Calculus I using group learning and a limited microcomputer laboratory component because of the restricted capabilities of the available microcomputer equipment.

After his sabbatical, he wrote proposals to the National Science Foundation to equip a microcomputer laboratory (ILI) and to fund a one-year trial implementation of calculus reform. The implementation proposal was funded but not the ILI grant. However, the University provided funds to equip the microcomputer laboratory. After the initial year's funding, subsequent NSF grants for a two-year implementation of calculus reform and an additional microcomputer laboratory were funded.

Oakland: The process began with a charge from the Chair of the department to the Committee on Undergraduate Programs to review the current language on multiple section courses, and to submit a revised section on policies for 100–200 level courses to the department for approval and inclusion in the departmental handbook. This charge, and the reasons for it, were discussed by the Chair in the second department meeting of the 1990 Fall semester. A deadline was given for the middle of the next semester.

Planning Initiated

At what stage did the department become involved in planning?

Cal Poly: Specific plans for change came in two parts: the first initiated within the department, and the second initiated a year later by our President. Shortly after the new pre-calculus course had begun, the calculus reform movement began to be mentioned more frequently in journals and at meetings, which increased the interest in alternative teaching and learning methods. In 1992–1993, the associate department chair organized a series of seminars in the Mathematics Department for mathematics faculty to share what they were doing in their mathematics courses with other members of the university. Members of the seminars also reviewed calculus reform materials and presented their opinions at the seminars.

At the end of the year, a department meeting was held to discuss a proposal to experiment with a calculus reform text in the next year. The meeting was attended by only half of the department, and some of those who wished to use the reform text thought of it as an individual experiment rather than as a major shift. Nevertheless, this was the beginning of departmental involvement. Issues that arose at this meeting include: the fairness of some sections of calculus using a different text in the middle of a budget crisis that decreased the number of sections offered; the fairness of requiring students to purchase calculators at a time when student fees were being increased yearly; and the use of assessment to measure the success of this approach.

In 1994, our President announced a grant program to assist departments in major changes in instruction. This led to a second series of discussions within the department about the need and desire for departmental change. The results of a survey and discussions with individual department members showed that a majority of the department wanted to add new techniques to their teaching style. However, they wanted instruction in these new techniques, and were reluctant to completely abandon the traditional calculus course. The Math Department proposed a four-year plan to train faculty in new teaching and learning methods. The goal was to train a sufficient number of faculty to teach all of the calculus sections.

Rio Hondo: In the fall of 1992, at our first department meeting, a calculus committee was formed. Its task was to develop a statement describing our philosophy regarding the calculus program. This was a good way to approach the whole subject—we didn't use the word change, we just requested a definitive guiding statement for our calculus program. The committee met twice a month and tried to accomplish specific tasks. For example, we each took a few schools in our area, called them and asked about their calculus programs. The members of the committee met on their own time. The only compensation was in the form of *flex* hours (each faculty member must accumulate approximately 50 hours per year in educationally-related pursuits).

Anon: I joined the committee early in 1991 as actual planning was getting under way. No additional people had significant involvement until the experimental offering was ready to go. The Preliminary Edition of Dick and Patton's *Calculus* was selected as the text. We liked the emphasis on reading, the de-emphasis on mimicking examples, the many thought-provoking questions, and the "rule of three" balance in the presentation. We envisioned much more interactive classrooms and much more thoughtful discussion following the increased student preparation through reading and attempts at problem solving.

Mount Holyoke: The discussions in 1986 here and among the other four of the Five Colleges (Amherst, Hampshire, Mount Holyoke and Smith Colleges, and the University of Massachusetts at Amherst) culminated in the successful Five College NSF proposal *Calculus in Context* in 1987.

Many of our faculty, more than just those who were writing, participated in regular discussions about calculus curriculum and pedagogy through 1987–1988 and the summer of 1988. Certain principles guided our discussions, principles gleaned from the experience at Hampshire: calculus is a language as well as a tool, and students should be able to read and write this language; differential equations should be fundamental objects of study; successive approximations lie at the heart of calculus; the study of modeling should include attention to qualitative as well as quantitative analyses; technology (for visualization and for computation) can and should change what we teach. Still, the discussions were messy and often contentious ("frank and comradely" was the phrase we used), and everyone's thinking evolved over time. For the most part, we achieved a strong degree of consensus for change.

Faculty in the mathematics-using disciplines at all of the Five Colleges were invited to give presentations to our working group and other interested faculty. The goal was for them to help us think about curriculum, but there was another advantage in that it informed colleagues in "client" disciplines about what we were doing at a very early stage of the process. One mistake we made was in not continuing these contacts enough throughout our work. Writers and consultants were supported by our NSF grant.

Ole Miss: Prior to the above activities, the Department had devoted considerable time and effort to an extensive review of its undergraduate curriculum over a period of four years. During that time, the Department developed a mission statement in order to set specific goals for the undergraduate program, and provide a method of assessing the department's progress toward attaining its goals. During this same period, the Department was a participant in a national project conducted by the American Association of Colleges and funded by the Fund to Improve Postsecondary Education (FIPSE). Under the auspices of this project, external examiners from Louisiana State University and the University of Alabama examined and interviewed graduating seniors at the University of Mississippi for two successive years.

One result of the Department's extensive curricular analysis was the confirmation that it was following a national trend of not meeting all of the desired goals in the teaching of calculus, even though the students' evaluations of these courses are generally high at the University of Mississippi. From its extensive research, the Department identified several noteworthy calculus projects developed at other universities. As a consequence, the Department decided that it should not develop its own materials, but should seek the project materials that had already been developed that match the needs and goals of the Department.

Oakland: In its initial discussion, the Committee decided that it would be important to structure a process that would be as open to as much faculty input as the faculty chose to provide. To that end, a survey was devised on current faculty attitudes; it was administered in an interview format by various members of the committee with individual members of the department. The Committee also decided to schedule an open hearing to review its first draft before a final draft was attempted. An overall outline of the proposed process that would be followed was communicated to the faculty.

Plan Development

Describe the actual planning process.

Cal Poly: It was decided at the meeting at the end of 1993 that in the 1993–1994 year there would be three sections using a calculus reform textbook. The students would be required to buy graphing calculators, but no brand would be specified. Students would be allowed to transfer out of the reform sequence to the traditional sequence, but not the reverse. In this way, there would be a group of students who had been through the entire sequence who could be used for assessment. Other, more general, decisions were made at that meeting. Only one experiment involving a different textbook could be conducted at any one time, so that it would be possible for students to schedule classes using the same text throughout a sequence. The schedule of classes must have a footnote by all courses that require calculators. Faculty members who feel it necessary to require a certain brand of calculator (for example, the HP–48SX for linear algebra) are responsible for having that specifically footnoted in the schedule. The right of faculty members to experiment with different pedagogy was re-asserted. However, instructors were encouraged to experiment in pairs for support and feedback.

At the end of the 1993–1994 academic year, the Math Department was awarded $10,000 for the first year of their four-year plan, with future funding dependent on progress. The first year was devoted to training a team of five faculty members in new teaching and learning techniques. The funds were used to send faculty to workshops, seminars, and conferences, to purchase software and books, and for release time for the sponsor of the grant. The five faculty experimented in their classroom with new techniques and reported back to the others. One member of the team joined the group of instructors teaching calculus reform sections. Minimal assessment was performed. The progress of students who had completed a year of calculus reform was followed, and attitudinal surveys were collected by two of the faculty.

In the second and third years of the grant, workshops will be held by the core team to train more members of the mathematics faculty. Funding has been requested for release time for the core team. They will be responsible for planning and running workshops, and writing a series of short papers with suggestion for the use of technology and ancillary materials in calculus courses.

The fourth year of the grant will be spent informally training any remaining interested faculty. Assessment will be continued and expanded during the period of the grant.

Rio Hondo: Around December of 1992, we, the committee, agreed on a statement of philosophy that was presented to the department. Looking at our statement of what we thought a calculus program should be, and comparing it to what we had, it was obvious that some changes would have to be made. We began by looking at programs, books, etc. By April, the committee had decided to go with the Harvard materials. We were enthusiastic and vocal and the department approved of our plan. Paul Moreland and I were teaching first-semester calculus in the fall, so we were sent back to Harvard for the workshop in the summer of 1993.

Anon: We recognized that some "teacher training" would be necessary to promulgate the new emphasis. Our plan was to

- provide intensive training over the summer between the experimental year and the full-blown year, and
- have weekly meetings of all involved faculty during the year.

Mount Holyoke: At Mount Holyoke, we began in the fall of 1988 with two pilot sections of the course produced by these discussions. Ken Hoffman of Hampshire College taught one, and Lester Senechal taught the other. Having an experienced instructor working in tandem with a (relative) novice worked extremely well. In the spring of 1989, my colleague Margaret Robinson (who had taught the previous year at Hampshire) taught another pilot section. Based on the success of the pilot sections, we decided to use the manuscript *Calculus in Context* text in all of our sections of Calculus I the following fall.

(In a subsequent semester, we experimented with an experienced faculty member in charge of the course, and two other faculty new to the revised course as "assistants." This actually did not work so well, partly because it was difficult to keep communications among the three complete enough to ensure "seamlessness," and partly because we are each used to more autonomy in our teaching. We had a similar unsuccessful experience with team-teaching with shared responsibilities.)

We measured success in three ways. The faculty were excited by and pleased with the new material. The students were enthusiastic and successful. (Interestingly, this was most noticeable in a spike in interest in minors in mathematics.) And faculty felt the students' conceptual grasp of the material was strong.

For the five years of our NSF project, all students in our reform calculus sections were surveyed in two ways: by the usual college teaching evaluation forms, and by special forms designed by the outside evaluator for our project (and used at all five colleges). Our evaluator also did interviews with groups of students at all the colleges. In addition, at the end of every semester, there was what we called a "debriefing session" for faculty at all the colleges who were teaching with the new materials (and many of our other department colleagues came as well). At the debriefing, we reviewed what had gone well and what had not, and discussed ways to improve the materials and/or the pedagogy. The debriefings were enormously valuable and resulted in many changes in the evolving text and in how we worked with students.

At Smith and Mount Holyoke we also arranged a sequence of half-day workshops for faculty in other departments (at Mount Holyoke we actually extended the invitation to all faculty, and had some positive response from faculty in humanities departments) in which we had them try some of the new things we were asking students to do, and discussed the goals of our new curriculum. We also invited their comments and suggestions. These, too, were enormously valuable, and in retrospect, I wish we had done even more. At Mount Holyoke we also had meetings with members of the physics and economics departments, our two principal clients. Again, with hindsight, we should have done more consulting at this level.

Ole Miss: Based on the Department's preliminary investigations, an existing calculus reform project was determined to be a suitable program for adaptation at the University of Mississippi. The University invited one of its developers to visit the campus for two days to explore more extensively the feasibility of the University's being a test site for the project. During that time, the developer had separate meetings with the Mathematics faculty, the chairs of the other science and engineering departments, the Deputy Director of the Computing Center and members of his staff, the Associate Vice Chancellor of Research and a member of the Office of Development, the Vice Chancellor for Academic Affairs, the Dean of the College of Liberal Arts, and the Director of the University's Writing Program. In addition, he made a presentation to the university community on his project.

After his visit, the Department's faculty agreed that the goals and the materials provided in this project are excellent and match well with the goals and needs of the Department. The communication with the department's client disciplines and the central administration established a framework within which calculus reform could be implemented with the complete knowledge and support of others outside the Department.

Oakland: It came as no great surprise when the survey revealed that faculty were much more amenable to telling graduate students and part-timers what to do than similarly telling other colleagues or themselves. In addition, there was also a distinction in attitudes towards freshman- and sophomore-level courses; at the sophomore level, the balance was heavily toward more faculty control rather than more departmental control. Consequently, the committee chose to accommodate these varying views by devising a policy which had three essential elements. The first was a statement of the objectives that the department had for these courses and the accompanying polices governing them. The second was a set of four different possible aspects of departmental control for a course. These included a "departmental" syllabus and course information document for students, a "department" information document for course instructors, a departmental final examination, and a course Leader, a faculty member assigned to be responsible for the course over several semesters. Initial specific combinations of these features for specific courses were proposed as part of the new policy. A third element was the specific assignment of responsibilities for making changes in an individual course and/or the total policy amongst the teaching faculty and the Committee. These elements were debated in the promised hearing, revisions made, and a proposal submitted to the department. After much discussion and amendments (friendly and otherwise, passed and failed), a final document was approved at the end of the Winter Semester of 1991.

Implementing the Plan

Describe your experiences as you made changes to your program.

Cal Poly: There have been some setbacks to our experiment with a calculus reform text. Some instructors became unhappy with the text and projected that unhappiness to the students. A variety of teaching techniques were used in connection with the text from traditional lecture to 50% of class time spent in groups, with mixed success. As a result, the sections have become unpopular.

It is hoped that this can be remedied in the next year. At least two more faculty are scheduled to join the team of calculus reform instructors, and both have indicated a desire to use alternative teaching and learning techniques.

An unexpected benefit of the move towards change resulted from a fear of some faculty that the calculus reform sections were not covering enough material. That fear led to the most productive discussion of calculus topics the department has had in years. The department agreed to delete several topics that had been taught in the traditional sections. The team of faculty leading the change decided that a common final test was a necessary assessment tool. However, difficulties are foreseen in obtaining agreement from the faculty to administer the test and the labor required to create and grade the test. Finally, the department has not heard yet about continued funding for the four-year plan.

Rio Hondo: In the fall of 1993, our first-semester calculus classes used the new materials, we required graphing calculators (TI–85), and we implemented collaborative learning. Quite a change from our traditional ways! Each semester we extended the program; all three semesters are now using new materials. We also got a motion passed by the department that states, "Only people who attend a workshop and agree to the new techniques will be scheduled to teach any of the calculus classes." Also, new people would have to begin with first-semester calculus. Since we only have 5 sections of calculus per semester, Paul and I covered all the classes for the first year. In the spring of 1994, I held a calculus workshop at my house, about 10 people attended, three of these were from Mt. San Antonio College (a nearby community college). We showed a video, talked about what we were doing in class, and then broke into groups and worked some problems.

Paul and I tried to get together and compare notes at least once every two weeks during that first year. We still do this, but not as often. Our biggest problem seemed to be getting the students to work well in groups. Our students were used to struggling alone in math, and they were hesitant to talk to each other. Also, we have a large number of foreign students that have marginal English skills. The first semester I had to force them to work in groups. I chose the groups, and they stayed with one group for five weeks. They received part of their grade based on group work. By second- and third-semester calculus, they didn't need any pushing; they had learned that there were real advantages in cooperation. I also have one "team" test; they don't know who they will be teamed with until the day of the test. Paul let his students choose their own groups; this didn't work very well. The groups tended to be based on friendship, ethnicity, or ability. And there were a couple of loners that no one wanted to include in a group. Writing tests has proved to be very time-consuming for me. The difficulty is designing a test that really tests for concepts, and one that acknowledges the use of a graphing calculator.

Anon: In our experimental offering we had 10 sections of Calc 1 in the fall of 1991. These students were compressed into 8 sections of Calc 2 for the spring. All of the instructors involved were positive about the program and worked together very well. The weekly meetings were productive and positive. But, alas, it is not the case that all went well. The students were not at all receptive to the thought-provoking questions. They clearly wanted to get to the rote as soon as possible. They did not want to understand—they wanted their formulas, their tests, their grades—they wanted to put math behind them as soon as possible. This was no revelation to us. However, the strength of their resistance to the new perspective was revelatory. They simply would not *think*. They really believed that ultimately they would be given something to regurgitate—and that regurgitation would result in a high grade. This was not the case, and yet another blow to morale occurred.

Early in the experiment we encountered another significant problem. The students were either unable or unwilling to read for technical substance. Oh, they could read aloud—and well. They could even read and glean facts. But, they seemed entirely unable to grasp a new idea from the written word. We had some success in improving their reading by providing specific reading questions to guide them. But, the reading problem alone nearly derailed our reform effort—so much of what we hoped to do was based on the unwitting assumption that the students already knew how to read (really read).

We had "bitten off more than we could chew." The combination of the attitude problem (motivational) with respect to thinking, the willingness/ability to read, and the ill-will spread by our students to their peers, persuaded us to cancel the full scale implementation. We are back to a traditional book, but making a careful effort to increase thinking and to emphasize reading. We hope these efforts will contribute to the students' overall education.

Mount Holyoke: It would almost be simpler to answer the question: What unexpected difficulties *didn't* arise? Some of our problems were the result of using new, evolving curricular materials that hadn't been adequately classroom-tested. Presumably those are not relevant for people considering adopting materials today. But others, I think, lie in wait even now.

We began with pilot sections of our new calculus, so students who registered for the pilot sections were choosing something new. Students in the pilot sections were very enthusiastic and almost uniformly positive. Reactions when all of our sections used the new materials were markedly different, and nearly half of our students were not only negative, but even hostile about being "experimented upon" (as they perceived it). Now, it must be said that there were many flaws in the early versions of the materials, and much justice in many of the students' complaints. But even after the materials had undergone many revisions and most of these difficulties were overcome, a minority of deeply unhappy students remained.

Recently, we have gone back to our old practice of letting the instructor choose the text for calculus, so there is diversity among our calculus sections. We had our most successful sections of *Calculus in Context* when we had a single section (taught by a sabbatical visitor) of traditional calculus. As far as we can tell, the difference was that students had a choice again—as they did when we offered our pilot sections. The difference was so marked that we are inclined to try to maintain something close to that degree of student choice in our offerings.

Ole Miss: During the initial implementation phase, almost all faculty taught at least one section of the reformed course. As the Department continued teaching sections of the reformed course, it became apparent that students, and to a degree faculty, were having difficulty with the expectation that the text should actually be read. This led the Department to adopt a different reform text while continuing with laboratory assignments and group projects. The Department has developed its own laboratory manual and uses group projects and microcomputer laboratory assignments, but continues to struggle with the selection of a text.

Oakland: The second phase of the project was in implementation. The portions dealing with departmental examinations and course leaders were done immediately. Next we developed instructor and student information sheets. The current course leaders were responsible for the initial drafts, and final drafts were submitted for departmental approval. These documents are the primary vehicles through which Department policies for specific courses are conveyed to instructors and students. Instructors receive theirs when they are assigned to the course; students receive theirs at the first class, after the course leader has filled in the specific items that are relevant for that semester. Since the electronic versions reside in the main Department offices, this can all be done quite easily, and facilitates operations at the start of each semester.

So what then have been the advantages and disadvantages of this entire exercise? The primary disadvantage is probably quite clear; it has been a very tedious process, requiring much patience from all of those who have been involved. However, on the positive side, there have been even more benefits than just those directly intended. Among the direct ones, there is now in place a well-defined procedure for how improvements may be made in introductory courses and made "permanent" beyond the faculty who introduced them. With the course leaders, we have increased the degree of faculty ownership for some courses and thereby increased the incentive for improving them. In addition, we have articulated criteria by which change will be considered and, once approved, evaluated. During the implementation process itself, a number of previous faculty initiatives in introductory courses were codified at the Department level.

Among benefits which were either indirect, merely hoped for, or quite unexpected, the following items should be included.

First, during the course of this exercise we have spent more time, as a total department, discussing specific instructional issues than at any time during the last ten years. We all have been involved in the discussion, not just an interested few. In this regard, the very process of debating how change could occur has made actual change happen. Trial balloons which were floated during the survey process, and received very mixed reactions, were adopted at the end with no opposition. Some examples would include uniform usage and encouragement of calculators and other computer support, sharing information on typical course grade distributions, and suggested distribution of problem types on tests and examinations. In addition, specific reform issues arose naturally (and indeed unintentionally) during the implementation process. The most significant of these led to the adoption of the Calculus Consortium at Harvard materials for all sections of the mainstream calculus. The SIAM/Linear Algebra Study Group recommendations for the first linear algebra course were ratified as well.

Second, the various documents which have been created in the process have been extremely useful in communicating with other units within Academic Affairs and more widely throughout the University. As we had desired, they have provided tangible examples of the seriousness with which the Department is pursuing its instructional responsibilities. Moreover, they have been received as such! Our colleagues in Student Affairs now regard us as the academic unit most concerned with actual student performance. University-level reviews within general education and enrollment management committees have also been positively influenced.

Third, our Chair is now in a better position to argue for increased budgetary support of various instructional efforts. This is because there is now a credible system in place for considering and implementing improvements in courses which are delivered by a large number of different instructors. Thus, money spent for potential improvements which prove successful will become money spent for lasting improvements. As an example, the necessary funding and cooperation to offer an experimental laboratory version of calculus went well during the previous academic year. In addition, one likely final result of the current effort underway with the business faculty will be a joint recommendation from both deans for additional instructional resources for the Department.

Advice

With perfect hindsight, what suggestions do you have for others planning to change their calculus programs?

Cal Poly: If members of the department are to take an interest in changing calculus instruction, they must feel that calculus instruction is important. Praise faculty activities concerning teaching at department meetings. Encourage discussions about current mathematics instruction issues. In addition, the Department should be kept informed of all policy shifts so that there is little reason to feel that changes are being made behind someone's back. Finally, it is not necessary to change everything at once, but don't underestimate your department either. They may be less resistant to change than is thought.

Rio Hondo: Spend substantial time clarifying the goals for your calculus program.

Our goals for our students are:

- to learn the fundamental concepts of Calculus;
- to think critically using the tools of Calculus;
- to foster and encourage curiosity for learning;
- to be empowered with confidence in their own abilities;
- to be prepared for successful transfer;
- to use technology as a tool not a crutch;
- to appreciate mathematics and enjoy its application in the "real world."

Our goals for ourselves are:

- to encourage professional growth and enthusiasm for teaching;
- to foster and encourage curiosity for learning;
- to exchange ideas;
- to expand our knowledge, versatility, and excitement about mathematics and technology.

Anon: Watch out for your hidden assumptions. Over the years we have learned to write tests which "justify" giving the grades we want to give. Our tests may or may not have tested what we really wanted the students to know. So they may or may not have actually learned what we have often convinced ourselves that they learned. Don't assume that the students want to learn! Many of them just want to fill the requirement. They have been cultured to obtain acceptable grades by last minute memorization and blind regurgitation. This, they have learned to put up with. But, press them to actually think and they will not put up with it. Don't assume a very high level of reading skill.

Please—try to reform. But go in with your eyes open. Test your local water on the issues that derailed me. Then set a realistic goal which improves on the *status quo*. Don't set your goals too idealistically. We all know what we want, but the magic word reform can't give it all to us in a single step.

Mount Holyoke:

1. Teacher preparation. It is hard to overstate how much things change when you move away from familiar territory. We have all learned over the years where students are likely to have difficulties and what their difficulties are likely to be. It is a real stretch to be in a situation where so many surprises lurk. Three things can help. Working with an experienced colleague is invaluable. Attending a "hands-on" workshop with the new materials can be almost as good, especially if you can be in e-mail or other frequent, easy contact with experienced users. A well-prepared Instructor's Guide can also be a great help if it points out pitfalls and difficulties to watch out for and offers suggestions for how to overcome them.

2. Working with technology. Teacher preparation is part of the issue here. It is essential, I think, for the instructor to do what the students are asked to do and with the technology they will use. One also has to be prepared for equipment to malfunction, either by having spares or by having a promise of priority attention from those who can help with technical problems.

3. Preparing assignments. It is easy to go overboard and ask students to do too much, partly in the enthusiasm of something new and partly because we don't have enough experience with different exercises. Particularly if you are using

non-routine exercises that make significant demands on students' reading and writing skills, you have to think about how much time that takes them. Grading such assignments also takes longer, and you have to plan for that time demand.

4. Using student assistants. Whether the potential assistants are graduate students or undergraduates, adequate training is essential. Especially in the early stages, you won't have a cadre of students experienced with the new materials. Inexperienced students can actually sabotage what you're trying to do. ("Let me show you an easy way to do that.") And student assistants who have been through a standard curriculum will not necessarily think your curricular changes are an improvement; their uncertainty or unhappiness can easily be picked up by your students.

5. Student stress. Virtually all the "reform calculus" curricula make more, not fewer, demands on students—and quite different demands than students are used to facing in mathematics classes. This makes students anxious. Many have relied on a single strategy in their high school mathematics classes: watch the teacher solve a problem, and then solve a set of similar problems by imitating the demonstrated methods. It can be hard for them to let go of a successful strategy and try something unknown, yet that is what we often ask them to do. Happily, some students are excited by the change and embrace it right away. Others, however, need a lot of encouragement and some serious "propagandizing" about why we make different demands and what they will gain from stretching in new ways. We have learned from our experience that it is essential to confront these student reactions directly and regularly in the classroom. We have learned, too, that we have to be very careful not to denigrate the skills and strategies the student has as we push him or her to move in new directions.

I would add just one more thing. Beware of deciding to incorporate technology before deciding what goals you wish to achieve and how technology can help you achieve them.

Ole Miss: Hindsight has indicated that the training component for faculty teaching the reformed course was significantly deficient. The primary lessons learned have been the need for:

- a "champion" for calculus reform within the department together with at least a small group of sympathetic faculty to assist;
- support of client disciplines and administrative offices;
- selection or design of a course that is compatible with the mission and goals of the department; (this assumes that the department has clearly articulated mission and goals);
- recognition of the variance of teaching styles and attitudes of the faculty;
- recognition of the variance of learning styles and attitudes of the students;
- adequate training of faculty for the proposed course; and
- an infinite amount of patience! This is a never-ending process.

Oakland: In summary then, my own personal advice on "institutionalizing" instructional change would be this. First of all, for your own sake, I hope you are not in a department which has so many people that need to be convinced before something really can happen! But whether you are or not, you will still have some people that will have to be convinced. In doing so, I suggest that you first try to get agreement on what your instructional goals really are, and what are the paths along which they will be pursued. What are you trying to do in your introductory courses? Who is responsible for proposing improvements? Who approves and makes "permanent" changes? Who evaluates the effect of changes? Specific agenda items will often follow more easily once this total framework of processes has been established. You may even get more change than you initially believed possible. Good luck, and remember throughout all of this to be open to other people's ideas. For reform both to work and to last they must be part of it.

Planning and Change: The Michigan Calculus Project

Morton Brown
UNIVERSITY OF MICHIGAN, ANN ARBOR

The Years Leading Up To Reform

At the University of Michigan, Calculus 1 has a fall enrollment of some 1900 students in 55 sections (class size about 35, mostly entering freshmen) meeting four times per week. Calculus 2 has a fall enrollment of about 1000 students, many of whom are also entering freshmen but with advanced placement credit. The instructors for these courses are also largely new to Michigan. They are new graduate student teaching assistants (TA's) or new Ph.D.'s. Many of these are foreign-educated. Only a very few senior faculty members had ever taught in this standard first-year calculus sequence.

For some years the Department has held a week-long training program for new TA's prior to the fall semester. In addition, the University requires all its international TA's to participate in an intensive three-week summer training workshop co-sponsored by the Center for Research on Learning and Teaching (CRLT) and the English Language Institute. Beginning junior faculty got little orientation; a single one-hour introductory session a day or two before classes began.

Often, the TAs' primary concern was (and is) with their own graduate programs, and the beginning faculty were (and are) deeply concerned with their research programs. Faculty instructors got a very small amount of undergraduate grader assistance. The TA's got no assistance with grading. Effectively, this meant that very little homework was collected and graded, and what was graded was seldom thoroughly corrected. The students' uniform exam grade largely determined their final grades. As a result, instructors often found themselves teaching to the uniform exams.

Before reform, a senior faculty mentor and a part-time TA designed the syllabus, coordinated the course, wrote the uniform mid-term and final exams, and met weekly for an hour with the 55 or so instructors of Calculus 1 to discuss course-related matters. The trainers (who had worked in the training program) and the mentors themselves visited classes on an "as needed" basis.

For several years preceding the reform, the Calculus 1 course (but not Calculus 2) progressively de-emphasized symbol manipulation and increased emphasis on concepts and geometrical visualization, all within the confines of a traditional text.

The Department operates a "Math Lab" which offers some self-study courses, but whose primary function is to provide tutoring for all elementary courses.

1991–1992: Early Pilots

In the fall 1991 term, we taught three pilot sections of Calculus 1 in which we introduced a number of innovations: (unstructured) team homework, graphing calculators (Texas Instruments loaned us 100 TI–81 calculators for the year), and a reform textbook in which we covered the same basic topics as the regular sections. At this stage, our thinking centered around the new technology, and the opportunity of the instructors to correct homework themselves. The team homework idea originally developed as a means of reducing the number of homework papers to be corrected. Only after the pilot began did we realize the powerful possibilities of team homework as a cooperative learning exercise.

Since a goal of the pilot was to anticipate eventual universal adoption, the first instructors were chosen to represent a spectrum of the teaching staff. One experienced TA, one junior, and one senior faculty member volunteered. Students were not told when they registered that they would be in a pilot section because we wanted a reasonable cross section of the student population. But we did not want students to feel like "guinea pigs" in a course which was as important to them as calculus, so they were allowed to switch to a traditional section later if they desired. In fact, only a few did so, primarily because they recognized that the pilot would be more demanding.

Our impression, by the end of the term, was that student performance on examinations did not seem much different from that in the traditional sections. An independent assessment by CRLT found that the students tended to be enthusiastic about both the calculators and the small group homework sessions involving their use. All the instructors found the course more interesting to teach, and felt that the students were more engaged.

The same three instructors tried repeating the pilot again (Calculus 1) in the winter term, but this time using the textbook used in the standard sections. There was much less success. One of the instructors reverted to a traditional course, and the other two could not reconcile the technology with the traditional text.

Spring/Summer 1992: Developing the New Program

During the winter and spring of 1992, we drew up a plan to completely revise first-year calculus in both content and in style. The plan was developed by the senior professor who had been involved with the program for the earlier pilot (i.e., a "zealot") and the senior lecturer who generally had the responsibility for the first-year calculus course. An instructional consultant from CRLT was involved from the beginning. The period from pilot to general adoption was to be three years. The Department, the College, and the NSF all committed to help set up the program. The chair supported the program enthusiastically. The attitude of the general math faculty initially, and at least for the next two years, was generally supportive but tempered by an appropriate "wait-and-see" attitude.

The College, the Department, and CRLT all contributed support for the initial stages of planning. We negotiated with Texas Instruments for them to donate 200 used TI-calculators and several view screens while the College paid (as matching funds to the NSF grant) for 200 more. We planned to loan these calculators to the students each term but give one to each instructor to keep. It would have been extremely difficult, if not impossible, to embark on this major innovation without these considerable extra resources. We anticipated that a great deal of support would be needed to help the instructors who were eventually to teach in this program. Also, we would have to help the students who would find themselves in a course different from what they had been expecting. We were planning a completely new course: new teaching and learning paradigms.

These are the main features of the program as it actually began in September of 1992.

- Incorporation of graphing calculators into the curriculum.
- A "reform" text (the decision of which book to adopt came just two months before classes began).
- Team homework; three or four students submitted one paper (our graders converted into Math Lab tutors).
- Cooperative learning in the classroom.
- Mid-semester assessment (the staff from CRLT visited each instructor's classroom around midterm, and then met with the instructor to give feedback and discuss what to change in response to students' suggestions).
- A short, but specialized, training program (the instructors were to all be experienced faculty and TA's) covering:
  - Use of the graphing calculator, how to help students get comfortable with it, and its pedagogical opportunities.
  - An introduction to the philosophy of the course and the new syllabus.
  - A workshop on classroom cooperative learning techniques using problems from the text as examples.
  - Some discussion of anticipated student reactions to the course.
  - A workshop on the CCH syllabus run by an invited Consortium member who had used the Harvard text.
- We prepared a pamphlet for instructors which commented on all the homework problems in the text and sorted them according to difficulty. (A

version of this idea was later incorporated into a publisher's instructor handbook relieving us of repeating this task for future textbook editions.)

We anticipated the more widespread changes to come by restructuring the Department's week-long TA training program. It was renamed "Instructor Development Program," and all the faculty who were new to Michigan were included in the same program as the incoming TA's (an immediate benefit of this was the social interaction between junior faculty and TA's). The new program emphasized student-focused teaching. The instructors who had volunteered to teach the reform sections were encouraged to attend those sessions relevant to the new course as well as the sessions developed especially for them.

1993: The New Program Begins

In the first semester we had ten sections. Three senior faculty, three junior faculty, and four experienced TA's all volunteered to teach the course (and, of course, attend the training program). The senior faculty all had strong research credentials. Furthermore, one had been an associate chair, and one was to become the new Department chair a few years later.

Students were not told when they registered that they would be in a section with a new format, and they were discouraged from changing sections to retain a valid comparison between the new course and the traditional course.

Our weekly meetings focused on issues of classroom management when one was using cooperative learning, team homework, student morale, and student understanding of the goals of the course. This last was particularly important. We had not fully anticipated the strong student reaction to the new pedagogy. (This kind of reaction is now well-known and characteristic of almost all calculus reform programs.) Fortunately, the instructors were quick to realize that we needed to devote some class time frequently during the first few weeks, and consistently during the semester, to discussions of the goals of the course, and reminding the students that they were learning how to think about math rather than just follow recipes. On the instructors' side, we definitely underestimated the need for morale building among ourselves, particularly the TA's, whose own pride required them to perform successfully in a style of teaching for which they had never seen a role model.

That first year, we learned a lot about using cooperative learning. For instance, we only began to introduce structure into the formation and activities of the "homework teams" after the third week of classes, i.e., we were learning on the job. The mid-semester feedback sessions carried out by CRLT helped us understand many of the frustrations and anxieties that the students were experiencing. Their fears were compounded by seeing peers in the traditional sections who seemed to be covering more material. However, the feedback also confirmed that the students were largely supportive of what was going on in the course. Naturally, there were many students who didn't buy into the program, and who criticized our changes. After a few complaints of this sort, any unsympathetic department chair or Dean might have quickly squelched the new program without a fair trial. Fortunately, we had full support from both our chairman and our Dean.

We made sure to document our new teaching styles. During the term each instructor was videotaped for at least one class hour. The videotapes were first viewed by the instructor for self-assessment, and some tapes were incorporated into future training programs and dissemination materials. (At this writing, some of the instructors who were taped are applying for jobs next year, and are using their tapes in the application process.)

As it turned out, the previous summer's extensive planning was largely successful. The year ended with a general feeling on the part of students and instructors (yes, we have evaluations) that the new program was quite successful. Students, with some reservations, were particularly enthusiastic about the positive aspects of the calculators and the team homework. During the term, to help with student morale, students were asked to name the program. Eventually the name "New Wave Calculus" was adopted. The student newspaper, after expressing reservations about New Wave Calculus in the fall, gave it a strong endorsement in the spring.

There were problems:

- the classrooms with strip tables or arm-chairs and a general teacher-in-front-of-the-classroom architecture were unsuitable for cooperative activities;
- the workload was much higher due in part to greater preparation time and homework grading, but mostly reflecting the emotional drain that occurs when instructors get to know their students as individuals, and care for them;
- the classes were too large for most of the instructors to get seminar type discussions going successfully.

It was still unclear whether such a program could be successfully instituted across the board. At the very least, we knew that a much more extensive instructor training and professional development program would be required.

Writing Our First Materials

An Instructor's Handbook was developed to assist instructors. The current handbook (1994–1995) familiarizes the instructor with the goals and features of the program. There is a section on student learning (how students learn), discussion of the setting up and monitoring of homework teams, cooperative learning, information about Math Lab and other program features, a section on student attitudes and typical problems associated with teaching, and detailed suggestions for each day of the first few weeks. Of course, there are also a daily syllabus and suggested individual and team homework assignments for the term. We also produced a brief (ten-page) Student Handbook, introducing the students to the new syllabus and pedagogy, and illustrating some examples of acceptable and unacceptable homework writing.

Setting Up the Instructor Development Program

Anticipating that there would be about sixteen instructors teaching twenty sections of New Wave Calculus 1 in the fall term, we felt that we should continue to keep the weekly staff meetings small, so we divided the instructors into two groups which were to meet weekly with a mentor. The mentors were a junior faculty member and a senior TA. Copying the cooperative pattern for the students, we formed self-support teams of four instructors who were to visit each other's classes, discuss teaching problems, prepare tests and class exercises, and commiserate with each other. This turned out to be sporadically successful. Some groups operated very productively. Most became dysfunctional.

1993–1994: Operating Traditional and New Wave Courses in Parallel

This year we ran two large calculus courses, traditional and New Wave, side by side. They used different texts and different teaching styles.

By now, the staff was getting stretched pretty thin; coordinating both courses, training the forty new instructors, and meeting them on a weekly basis. The operation was becoming too large and complicated. We assigned jobs so that within the New Wave, we had one senior faculty to coordinate the sections, while a TA and junior faculty "mentored." The traditional sections were coordinated and mentored by the senior lecturer who *also* taught a New Wave section.

It quickly became impossible to keep the courses as distinct as they had been in the previous year. We slowly began to add components from the New Wave to the traditional calculus course. Starting in the fall of 1993, all students in all the introductory calculus courses were required to buy calculators, and in the winter of 1994, we went across the board with the reform textbook. Additionally, the director of the Math Lab had revised the pre-calculus program over the summer so that the twenty fall sections began using the TI–82 and a reform pre-calculus text which emphasized more writing, discussions, and problem solving. We were also seeing some changes in the third-semester calculus which reflected the same general spirit. (These changes were being developed by a different cohort of faculty.)

In the New Wave calculus program the emphasis had refocused from technology issues to teaching and learning issues. The discussions in the regular mentor meetings developed our ideas. Near the end of the first semester, CRLT held a focus session for the New Wave instructors to get some feedback from them. This session was not only informative for us in preparing to go across the board, but helped the instructors focus on important issues concerning their own attitudes and needs.

All of us were concerned that there might be a weakening of the "basic skills" normally developed in Calculus 1, e.g., differentiating functions, so we agreed to have a number of such questions appear on both the traditional and New Wave final exams for the fall term. An analysis of performance showed little difference between the two groups, with the New Wave students scoring slightly higher. This made us more comfortable about continuing to make changes.

Beginning in the winter term, we switched all sections to the reform textbook. This meant that the only remaining differences from the student point of view were that not all sections used cooperative learning or team homework. We noticed that using a single text for all the sections made the students much less anxious about whether they were covering the "right" material. We found, however, that most of the instructors in the "traditional" sections spontaneously tried the cooperative learning techniques from the New Wave. It was clear that these instructors needed much more guidance than they had gotten in their first few days of training. We began developing far more specific instructional guides, and we had one of our experienced instructors videotaped each day during the first week of classes. These tapes were to be used in later terms to help instructors set the tone of the course.

The second year proved to be another successful one for the program. Once again the instructors and students alike felt they were getting a better course.

Spring/Summer 1994: Getting Ready to Take the Plunge

Our plans were to complete the changeover. We did not have the resources to direct two types of Calculus 1 and two types of Calculus 2 in each of the fall and winter term, so we went across the board with the new program. In the coming fall we decided to have all the sections of both Calculus 1 and Calculus 2 operating in the new way. (This did not include 10 sections of Calculus 2 which were reserved for advanced placement students who wanted to learn *Maple*.) We also began referring to the course as "Michigan Calculus" to avoid the connotation for students that we were trying something "new" on them.

Total Training

It appeared that we would have to train about 50 souls in the fall: new junior faculty, new TA's, and graduate students who were not new but had not taught this brand of calculus before. Some instructors would be teaching in the ideal setup; overhead projectors (and view screens), classes of 24 students with the 3' x 3' tables which made group work in class easy, but others would still be looking at inappropriate classrooms with 32 students. We decided that everybody should be trained in the new methods. The beginning-of-term training program, which now encompassed the seven working days before classes began, again seemed successful. This impression was later confirmed in several focus sessions held with instructors during the following winter.

1994 Fall: A Rocky Road

In the 1994 fall term there were eighty sections of Calculus 1 and Calculus 2, and the vast majority of the instructors had not taught before in the program. This once-only situation resulted from going across the board for the first time. We saw the immediate need for continuing instructor development. Several of the new junior faculty had little or no teaching experience. Furthermore, a number of faculty had brought with them attitudes about teaching that reflected the values of mathematicians in countries that have a very different prevalent philosophy from what we were trying in our program. Also, even in ordinary years, a percentage of our new TA's and faculty need a great deal of extra attention. Since this was the first time that the new program was not voluntary for the instructors, there were some (actually not many) who did not "buy into" its philosophy or want to put in the extra effort the new program required of its first-time teachers.

To run the fifty-nine sections of Calculus 1, we divided the instructors into six groups and had each group meet weekly with one or three mentors. The mentors themselves met regularly, and tried to coordinate their activities. Syllabus matters were dealt with by e-mail, so there were few large required "staff meetings." The instructors clearly would have benefited from some teaching "clinics," but we were hesitant to increase the time commitment for the instructors who were already being asked to do more than they had done (or seen their peers do) in previous terms.

With so many sections, the experienced CRLT staff person who had been doing the mid-semester feedback sessions in previous terms, now needed to have a new group of assistants. Thus, there was an extra layer of bureaucracy that slowed and somewhat distorted information in both directions between the directors of the program, the instructors, and the students.

As it turned out, because of extensive construction on campus, a very large number of the classrooms were dramatically unsuitable for any kind of teaching and frequently much too small for the number of students. Finally, to compound the logistical problems, the overall university enrollment was larger than expected, so quite a few classes remained at size 35.

Was the new program a success under these somewhat difficult conditions? Certainly there was less cooperative learning going on in the classes than we had hoped for, but still there was quite a lot. Many of the instructors did "buy into" the program. Some were very successful, many were struggling with the new teaching style, and many needed a lot more help than they got. The student mid-semester feedback, the instructor focus sessions, and the student end-of-term evaluation forms suggest that the students generally liked the team homework, although many expressed concern that their grades were being affected by the work of others. This third year of the program brought a downturn in students' evaluation of the course. The single number expressing "this was an excellent course" on their evaluation forms fell by about a third of a point on a scale of one to five. This drop can be partially explained by our problems with logistics in going across the board.

1991–1995: Logistics In Hindsight

We did not foresee that the amount of planning we would need would be about five times what we had expected. The centralized University structure made even small changes quite difficult. For example, getting the College to dedicate two classrooms for the program was a major endeavor. Then we had to get the rooms fitted with locked cabinets to hold the overhead projectors and the calculator demo view screens. There were lots of details, for example, even low-tech equipment needs attention: the demo calculator may need repair (or even new batteries).

In the summer of 1992 alone, we had to

- get the rooms ready,
- be sure the textbook would be published and delivered on time (the actual textbook choice was not made until the end of June, and the book was, at that time, in a preliminary edition),
- arrange for enough calculators for both students and instructors,
- explain our program to the College academic counselors, and to the counselors in the many other departments and colleges,
- design a more elaborate evaluation program,
- prepare the training program for instructors and tutors,
- plan, write, publish, and distribute Instructor Handbooks,
- plan calculator clinics for the beginning of the term.

Each of these jobs proved more complicated than we had imagined. Take, for instance, the evaluation program. The effort of hiring a principal evaluator, setting up the evaluation program, and then budgeting it, overseeing the budget, and coordinating this budget with the budgets of both the College (which helped fund the evaluation) and CRT turned out to be a major headache. During the 1993–1994 year we met with the evaluation team for about 3 hours twice a month, and it actually took about two years to get the logistics of the evaluation running smoothly.

If the project operates on a large scale, even simple problems can be time-consuming to resolve.

Observations About Teachers and Teaching in a Reform Course

1. For many instructors, a fully-invested teaching effort requires so much intellectual and emotional energy that it interferes with their graduate, postgraduate, or research programs. Certainly first-year graduate students should not usually be given primary responsibility for teaching a calculus course. They could apprentice, work in a Math Lab, or perhaps have supervised recitation section responsibility.

2. Be sure that the people in charge of the curriculum teach the course.

3. Be sure to attend to instructor morale. Everyone will need frequent reminders of the goals of the course.

4. Students will take responsibility for their learning if

   (a) they are given the opportunity, and

   (b) they are helped in learning how to take this responsibility.

   The teacher's responsibility is to help the students learn and help the students learn how to learn.

5. Since many students will not willingly accept the new ways and demands of reform calculus, they will want the "old way" back. Often instructors will use this attitude to justify their own temptation to return to a traditional teaching style.

6. We need to think hard about the differences between what we (or the text) say and what the students (differentially) hear. For example, many of the concepts that instructors may think of as "trivial" are perennial stumbling blocks for students.

7. Students can have ideas (correct and incorrect) about solutions to first-year calculus problems that we could never have guessed. This actually explains a lot of apparently nonsensical (to us) statements that come from reasonable students.

Observations About the Student Response to Reform

1. The students need constant encouragement. This is a big change.

2. Most students like the strengths of cooperative learning, but many are concerned about its perceived downsides (particularly in its relation to grades and scheduling problems).

3. All tutoring should be restructured so that the tutors learn how to guide students in problem solving rather than simply answering the team homework problems. Our undergraduate tutors usually could not solve the problems anyway which led to many complaints from our students, e.g., "How can we be expected to answer homework problems when the tutors can't even do them?"

4. Expect student evaluations generally to go up during the pilot stage, and then generally go down for a few years when you go across the board.

The Politics of Reform

1. Reform efforts must be accompanied by evaluation, and evaluation requires the setting of **goals**.

2. Get the client departments on board. Many are concerned with the same pedagogical issues, don't know what you are doing, and are often unfamiliar even with your old program. They can be a great source of support with the higher administration, and within your department.

3. Get the academic counselors behind you. Explain the goals of the course so that when disgruntled students come to them, the counselors can help defuse the situation.

4. Pasting on technology is not calculus reform, but it can be a doorway to reform. Practically, without an ingredient of technological innovation, administrators and funding agencies are not likely to support a reform effort.

5. The reform textbook (i.e., the content of calculus reform) is an instrument of reform; it is not reform in itself. The real reform will be in how we teach and how students learn.

6. We had difficulty getting senior faculty to volunteer to teach the New Wave course. In the first two years about eight senior faculty did so. On the other hand, that's a lot more than did so before the new program began. Also, as part of its bargain with the College, the Department pledged that a far greater number of senior faculty would teach a first-year course.

7. Because calculus (and pre-calculus) is such a large part of the teaching role of a Mathematics Department, serious reform will affect the whole operation of the department: hiring of faculty, support of TA's, the reward system, prioritizing course assignments, and the involvement of senior faculty in the elementary mathematics courses. Reforming calculus will require a considerable increase in resources. This must be done through an increase in resources for the department as a whole and a reprioritization of resources within the department.

Planning and Change: Some Things Need Time

As the introduction to this section on planning and change said, "For each department, the development of a plan for change will depend primarily on the talent and energy of the department members who support change, and the general willingness of the community to consider and participate in the redesign of their program." This essay suggests some directions for the planning process, and ends with the checklist that was mentioned at the beginning of Part II.

Stages of Change and Planning

As the reports from different schools illustrate, both change and planning for change take time and energy. There are five common stages in the planning process that we will discuss.

1. The pre-planning stage: what happens before the department is involved.

2. The initial planning stage: a departmental process is organized, background information is assembled, values and goals are clarified, and preliminary visions are formed.

3. The plan development and adoption stage: goals are adopted, objectives are developed based on the goals, and plans are formulated and adopted for both transitional and final programmatic changes.

4. The transition stage: implementing the planned changes begins with faculty development and training; new materials, teaching techniques, technology, and organization are phased into use; feedback is gathered and evaluated in terms of goals and objectives; and plans are revised as appropriate prior to final changes.

5. The full implementation stage: all final changes are implemented; training and evaluation continue; revisions and additional changes are developed.

These stages are not disjoint, and their intersection may be quite substantial depending in part on the scope of the changes envisioned and the depth to which any particular issue is pursued.

Pre-Planning

This is a time when some faculty begin to see a need for change in the way they teach. During this first stage of planning, informal groups form to discuss problems in the current operations, and share desires to improve instruction by a variety of means. This stage may last for months or even years. During this period you will observe individual initiatives and actions such as attending workshops, developing personal supplementary materials, using alternative forms of assessment, and introducing experimental learning situations to individual classes. Generally this stage involves changes in individual behavior, and motivates more lasting and global changes.

Initial Planning

Eventually, discussion of change becomes sufficiently acceptable, or some outside agent like a dean intercedes and the department as a whole becomes involved in a more formal initiative for change. At this point some key decisions on the role and depth of planning are made. To ensure that a planned approach is followed, it is important at this stage that the planning process be discussed as much as, or perhaps even more than, initial visions of the scope and substance of proposed changes.

Key elements of the actual process can vary widely from school to school, but in designing the process it is important at the initial stage to open communication with as many as possible of those affected by the calculus program. Broad and open participation in the planning process will bring fresh ideas and the creativity of the college community to the problems. The groups involved should include faculty and staff from other departments and students at differing levels of mathematical experience. A collaborative effort in designing and planning change will empower those who participate and helps spread the support for the change beyond the initial promoters. Committees can be organized or individuals can be selected to research and report, to learn and demonstrate, and finally to assure that people have a sense of the possibilities for change before making decisions. Without a good understanding of the nature of the possible changes, the final design may be less than optimal. So in this early stage, besides getting people involved, it is important to stay in a fact finding mode and not to jump too quickly to conclusions. Examining alternatives at this stage is more than just looking at a couple of new wave books and two or three different pieces of technology. Go to conferences; talk to people at other schools about their experiences; listen to people reporting on their own successes and failures; seek out others using materials you are considering for direct feedback.

As the collecting and involvement process goes forward, departmental sessions should focus on recognizing goals and considering benefits, weaknesses, and costs of alternative changes. Aim for consensus on a vision of a program that will make change. A fracture in the department's willingness to work together on change can lead to dissent that will haunt and disrupt the process beyond the value of holding onto any one specific item. Try to put personal agendas on hold and focus on finding common directions. Avoid a tone of finality. Nurture the concept of change as an ongoing process, so that time will be on your side.

Plan Development and Adoption

After working through the discovery process of initial planning, it is time to fashion a proposal for initiating change. At this stage you will need to keep the long-range vision distinct from the short-term attempts to reach for the goals you have articulated. Don't forget to plan how to assess whether or not the changes are achieving your goals. The pace of change is sometimes a more important factor in the resulting success or failure than the substance of the change. Keep in mind that both faculty and students will have to make changes in the way they teach and learn. This means planning for faculty development as well as training other academic support personnel such as tutors, graders, and teaching assistants. And keep in mind that students during the transition period may not have as much support from classmates who have been through a different educational route.

After the transition and tentative final changes have been outlined, discussed widely and thoroughly, and agreed on, the planning doesn't stop. There are many details to watch, much communication and preparation to be continued, small and large problems to resolve, and new changes to consider that were not foreseen during the planning stage. It is important to think of the planning as an on-going process to ensure that problems are expected during implementation and not treated as catastrophic. Expecting problems will reduce anxiety when they arise. Having a mechanism to deal with difficulties will prove to be prudent.

Transition

As you come through the transition stage of implementing your plan, you may find discouragement and dissolution along with some unexpected elation. Making changes will be even more demanding than the planning, and this will not diminish for some time. Plan a supporting system to keep the positive parts of the process you developed during the planning stages active while implementing your next changes. This will help you avoid some of the regressive tendencies that come with fatigue.

Full Implementation

When you come to a point when you think you are done, think again. There's always more change in the future. With the planning experience gained from changing your calculus program, you can do more with your other programs. Keep rethinking your goals and reviewing how your students are responding to your program. As people and the educational environment continue to change, your program will need to adapt to stay effective and lively.

Conclusions

Party plans and the development of a plan for changing a calculus program do have much in common. They both take time and careful attention. The risks of not planning a party are only temporary and if the party does not succeed, it is unlikely that any serious harm will result. The same might be said about previous changes in the calculus programs that focused primarily on rearranging the order and selection of topics and a textbook adoption. The changes being discussed now for the next century have a greater scope than any reform in the past fifty years, and they carry with them a correspondingly larger risk. It must be recognized, however, that there are also serious risks in not changing; there is much evidence that traditional methods have not been as effective as we like to think. This much is certain, whether one opts for change or maintaining the *status quo*, the results are likely to be much more satisfactory if they are planned. There is no guarantee that planning will bring success to changes. However, even if the changes don't bring total euphoria, the benefits of planning creatively and carefully lie in the way that the people involved in the process can grow and continue to engage in the process. The possibilities for success from planning change are abundant, and in some sense you will gain even if you fail. Is there any reasonable choice but to plan?

Checklist for Planning Change in a Calculus Program

I. Pre-Planning Stage: Individuals starting change.

  A. Who goes first?
    1. Individuals or *ad hoc* informal groups (a grass roots development).
    2. The department chair.
    3. The dean or other administrator from outside the department.
  B. Be prepared: What response/environment will you encounter initially?
    1. Who will be likely to be cooperative or supportive?
    2. From what directions should you anticipate resistance?
  C. First images: What kind of changes do you envision initially for yourself?
    1. Individual instructors use technology?
    2. Different types of assessment?
    3. Classroom dynamics altered?
    4. Experimental section(s) offered?
    5. Textbooks and supplements?
    6. Laboratory experiments and/or projects?
  D. Take stock of the personal commitment to change in your college community.

II. Initial Stage: The department becomes involved.

  A. Formulate procedures to develop a plan.
    1. Open discussion of a vision of your calculus program in the future.
    2. Form a timeline for articulating goals and objectives.
    3. Consider the kind of support (e.g., release time, retreats, ...) you need for the planning.
  B. Survey the situation and environment.
    1. Make a list of your resources. Be sure to include all the school's physical resources that might be useful, not just those the department happens to control.
    2. Make a separate human resource list with faculty, staff, and student resources based on experience and potential.
  C. Investigate what other departments and practicing professionals need and want from your courses.
    1. Make contact with the other departments personally.
    2. Encourage discussions by other departments of how they use calculus in their courses and how professionals use calculus in their work.
    3. What higher level thinking skills do other departments want?
    4. Do other departments see the calculus course providing more than a list of technical/mechanical results and skills?

D. Investigate what your students want and expect from a calculus course.
   1. What are your students' learning and studying assumptions?
   2. What do your students assume about assessment?
   3. What do your students expect to learn /achieve?

E. Investigate the organizational structures that control your program and change.
   1. What are the calendar assumptions?
   2. What are the class/course structure assumptions?
   3. What are the curriculum constraints?
   4. Who will need to be involved in approving changes inside and outside of the department?

F. Organize for change and foster communication.
   1. Will your department organization provide effective leadership and support to facilitate communication and manage resistance?
   2. What opportunities have you created for cooperation, open discussion, and creative thinking?
   3. Will your organization provide separate mechanisms for dispute resolution and decision making?
   4. Have you established input and review connections with other parts of your academic community?
   5. Have you contacted appropriate administrators to involve them, gain support for the planning process, and open some dialogue on the costs of change?

G. Developing a vision based on needs, desires, and resources.
   1. Have you considered the needs of students and the capabilities of the faculty and staff?
   2. Have you set goals that are sufficiently realistic and ambitious?
   3. Have you considered what changes will be needed in resources to implement your plan?

H. Develop a plan that responds to a vision.
   1. What is the initial vision for this plan?
   2. How extensive is your original vision of change?

III. The Plan Development and Adoption Stage.

A. Who will be involved and how?
   1. Form a committee.
   2. Consider how its membership is determined.
   3. Who to consult: client departments? students?
   4. How frequently?

B. Does the plan focus on a single primary change and related secondary changes or take a balanced view in changing several parts of the course simultaneously? Is diversity or monolithic change the final objective?

C. Is your time frame reasonable for (phasing in) the planned changes?

D. Have you planned professional development to support the change for those actually teaching and providing support services in the reformed program?

E. What provisions were made in the planning for flexibility and revisions?

F. Who will pay the costs of the changes?

IV. Implementing The Plan.

A. How will unexpected difficulties be handled for large issues?

B. Do you have a mechanism for resolving future questions?

C. What kind of review and assessment have you planned?

D. What parts of the plan will be more or less changeable?

Part III:

Assessment

The goal of this section is to highlight methods of evaluating student learning and student work in a modern calculus course. In the article *Assessment*, David M. Bressoud discusses a variety of techniques that he uses in his classes to judge students' progress. He stresses that assessment must emerge from clearly articulated goals and must be an integral part of the course. Information gathered in the assessment process should be used to find ways for improving the teaching of the course as well as the performance of the students.

Assessment is also tied to the process of assigning grades, and the most commonly used tools for computing and assigning grades are the results of tests and quizzes, and final examinations. The importance of the final examination is somewhat diminished in courses where homework assignments, laboratory reports, and writing assignments all contribute to the course grade, but most calculus courses still conclude with a final examination that is taken by all students who complete the course.

Following the lead article *Assessment* is a collection of final examinations given in 1994–1995 to students in calculus courses. Almost ten years ago, a similar collection of final examinations was assembled for *Calculus for a New Century: A Pump Not a Filter* (MAA Notes, Number 8). Whether or not one subscribes to the idea that a final examination reveals what has been going on in a course, it is interesting to compare the current collection to the one from ten years ago.

The final examinations were gathered from large public universities and small private colleges, from Ph.D.-granting universities and two-year colleges, from a diverse group of schools where the teaching strategies and teaching materials range from very traditional to very *reformed*. The texts of the examinations are reproduced here exactly as they were presented to the students. Occasionally, we made minor changes in some graphs when it became difficult to reproduce a hand-drawn graph in electronic form. Information about the points awarded for a problem has been omitted, but all relevant information about the use of technology or allotment of time has been included.

Changes in the teaching of calculus, the use of technology, and the increased emphasis on concepts are also affecting two important instruments of assessment, the Advanced Placement Examinations (AP exams) and the Graduate Record Examinations. The Advanced Placement calculus curriculum is currently under revision, and, as stated in the Part I article, *Visions of Calculus*, "the new AP calculus syllabus will be a measure of the institutionalization of modern calculus courses." To provide information on the changes in the AP exams, we include at the end of this section an article by Ray Cannon and Anita Solow that describes the changes in the AP exams.

Assessing Student Performance

David M. Bressoud
MACALESTER COLLEGE

The biggest problem with being a student is that you're always too busy getting an education to learn anything.

—attributed to Richard Feynman [12]

Introduction

Richard Feynman has put his finger on the reason why it is so difficult to get mathematicians to take assessment seriously. Most of us recognize the educational ideal to which he is alluding: the teacher should share hard won insights, signal traps, coach, challenge, and encourage while the student struggles, stumbles and rises, and—by dint of hard work—comes to a personal understanding of the subject. This is what learning should be, and within this scenario, testing seems at best a distraction.

We do use tests of knowledge, skill, and understanding. We know that we must, not only because our institutions insist upon it, but for our own peace of mind. Whatever our ideal may be, we live in an imperfect world in which such certification is the motivating force behind much of the learning that takes place. We bemoan the perennial question, "Will this be on the test?", but we know thatthis question is inevitable and even rational given our educational system and the forces that send students to us to learn mathematics.

I begin with the assumption that I will always have students who will not study without such external motivation, and that I have an obligation to teach them. I have found that most of these students want to learn, that they do want to understand, but that there are limits to their single-mindedness of purpose, to their willingness to face repeated failures in the quest to understand. I have also found that almost all of them carry a spark which, if nurtured and encouraged, can burst into the flame of a passion to learn.

Assessment constitutes one of the bags of tools at my disposal. It is not a recent discovery that it contains more than tests and can accomplish far more than certification. There are assessment techniques that can help me keep my finger on the pulse of the class so that I can react when confusion is spreading. There are tools to help a student recognize misconceptions in time to correct them, and confirm when personal progress has been made. Assessment can enable the student to integrate seemingly disparate topics and can provide challenging and summative learning experiences. These tools are not new. Good teachers have used them for generations. But not all of them are widely known or widely used.

As we proceed with the task of reinvigorating our teaching of calculus, we need all the help we can get. The purpose of this article is to display some of the tools of assessment that others have found useful, and to challenge each of us to think about what assessment means and how we want to use it. My own paradoxical experience is that when it is used well and wisely, it can actually move me and my students closer to that ideal in which testing has no place.

This is necessarily a very personal and anecdotal treatment of assessment. I hope that what I have learned may be useful to others. I will also dare to make general statements about assessment. In a few instances they will be supported by research or statistics, but in the spirit of John Ewing's comments on assessment in the *Monthly* [6], I am relying primarily on presenting a case that will make sense.

Basic Principles

Much of the current work on assessment involves the entire undergraduate program in mathematics and asks what the students are learning and how well they are able to use what they learn. This is the focus of the assessment report of the MAA's Committee on the Undergraduate Program in Mathematics [5]. We shall look at these questions within a much narrower context: the individual calculus course. Within a specific class, assessment is intimately tied to evaluation and the process of assigning grades. This is appropriate, but we need to be aware of the wider uses of assessment even here.

As we look for more effective ways of assessing what is happening in our classes, there are three important principles:

1. What is assessed and how it is assessed must be seen by both faculty and students as emerging from clearly articulated goals and objectives.

2. The assessment must be an integral part of the course.

3. The information gathered from assessment must be used to improve both our own teaching and student performance.

Goals and Objectives

Calculus tests are usually constructed by looking over the syllabus for the past few weeks, determining what knowledge the students should have picked up, and finding problems that use this knowledge. The first calculus test invariably includes the obligatory derivative of a polynomial (to build confidence) followed by problems that use an assortment of techniques and applications of differentiation. We are assessing the students' ability to use this knowledge, and the message that we send is that the mastery of the techniques of differentiation—with some sense of how to apply them to standard problems—has been the objective of the first weeks of class.

If we actually ask students and faculty what students should be learning, the answer goes beyond manipulative skill and well-practiced applications. Both students and faculty use the word "understanding" when describing the objectives of a calculus course. It is useful to take this concept apart so that we can get a better handle on what it would mean to evaluate understanding. Bloom's taxonomy [3] of the major categories of cognition is helpful:

1. **Knowledge:** the ability to remember previously learned material.

2. **Comprehension:** the ability to grasp the meaning of material. This can be demonstrated by translating (for example between graphical or numerical or symbolic representations) or by explaining or summarizing.

3. **Application:** the ability to use learned material in new and concrete situations.

4. **Analysis:** the ability to break the material down into its constituent parts so that its structure may be understood.

5. **Synthesis:** the ability to recombine constituent parts into something that is new.

6. **Evaluation:** the ability to judge the appropriateness of using what has been learned for a particular purpose.

True understanding involves all of these. If we value more than knowledge, then we must test more than what can be memorized.

Our first task is to establish clear goals for the students who will be taking our course. Bloom's taxonomy is generic. We must translate these categories into specific goals that are appropriate for calculus. An example of such a set of specific objectives for first-semester calculus is the one developed by Smith and Moore for the students in *Project CALC* [9]: "Specific goals for this semester are for you to:

1. Understand the concept of a function in a variety of representations.

2. Understand the concept of a derivative and its relationship to rates of change including linear approximations, Newton's method, Euler's method, tangent lines, and instantaneous velocity.

3. Formulate problems involving rates of change as initial value problems.

4. Solve initial-value problems both numerically and formally, and be able to explain and use your solutions.

5. Develop familiarity and facility with a computer as a tool for understanding mathematics and for solving mathematical problems.

6. Write reports using a word processor.

7. Differentiate functions using the standard techniques of differentiation.

8. Explain the relationship between data and theoretical models as a means of examining real world phenomena."

Whatever others may think of this particular set of goals, the point is that they go well beyond what we usually test and they give definition to what it will mean to understand calculus. There is no universal set of goals for any calculus course. Each department must determine objectives that are consistent with the nature and expectations of the institution and the preparation of its students. The individual instructor then has the responsibility to flesh these out, keeping in mind personal strengths and preferences and any special circumstances of the students that he or she will be facing. It is important that both we and our students know where this course is going and what is expected.

Making Assessment Part of the Course

A wonderful set of goals is useless unless students are held accountable for achieving them. Most students come into a calculus class with a desire to learn and understand, but few have a passion for learning. Many things compete for their time and attention, and most students will do the minimum of work that is required to attain what they see as a satisfactory grade.

If your course objectives include the ability to summarize and explain a theorem, or to analyze its proof, or to clearly and cogently lay out the reasoning behind the solution to a problem, or to choose an appropriate technique to explore and solve a new and unfamiliar problem, then you must test these abilities. Our standard, timed examinations are not sufficient for this purpose. Few students can demonstrate higher-level thinking skills when they are under time pressure. Such examinations must be supplemented by other assessment tools such as take-home exams, projects, reports, and oral presentations.

If you want the student to do assigned homework problems, then collect and grade them. If you want the student to learn to read a textbook and digest the main points before you give your presentation, then give periodic and unannounced quizzes on the material they were to read. If you want the student to master a basic collection of techniques of differentiation, then test these techniques and insist on perfection (see section 3.1). If you want the student to be able to write clearly, then collect, critique, and grade written exposition.

Using Assessment to Improve Teaching and Learning

The third principle is critical. How often have we constructed a final examination that we feel really gets to the core of what we want our students to be able to do, only to find that the entire class performs abysmally? We have put ourselves in the untenable position of failing everyone or of giving A's and B's that we know do not mean what we want them to mean. I myself plead guilty to having revised the goals for a course in the light of the results of the final exam.

The point of the third principle is that there should be no such surprises at the end of the semester. This should not be the first time that we evaluate comprehension or the ability to apply, analyze, or synthesize, that we ask students to be creative and insightful. Our expectations of what they are to achieve must be made clear from the very beginning of the course, and we must use the results of early assessment to help the students move toward the goals.

The need for early assessment as part of the teaching process is especially critical in calculus or any first-year course. Many students are not accustomed to analyzing, explaining, or applying what they have learned to new and unfamiliar settings. If all we do is say that they are now responsible for these higher levels of cognition, test them and find them wanting, and then throw up our hands and walk away, then we are abnegating our responsibilities as teachers. It is our duty to help them achieve the course objectives.

This can be done by front loading the course with many projects and challenging problems at increasing levels of difficulty. We must set aside the time to critique student efforts, to use their work as a springboard for discussing expectations, to show examples of what their classmates have done that is praiseworthy. It is especially important that students have the opportunity to revise and rewrite their early work. Assessment is not just for our benefit. We must use it to help our students recognize where their level of understanding is falling short of what we expect of them and what they need to do to correct this.

One example of how this can work comes from my own experience in first-year calculus. The problem was to find population, P, as function of time, t, given actual population data and the differential equation that P is assumed to satisfy. One pair of students found a correct solution, but their explanation of how they found it revealed a lack of understanding. They described the problem as: "to get $dP/dt = kP^{1+r}$ in the form of $P^r = 1/rk(T-t)$ [where] $T = 1/krP_0^r$."

This suggested that they were not solving a differential equation, they were performing a series of manipulations. This impression was confirmed when they used the data to approximate the value of r. They correctly recognized that $\ln P$ is a linear function of $\ln(T - t)$, but their explanation revealed that they had no understanding of why this was true: "For any function $P(t)$, the graph of $\ln P$ versus $\ln(T - t)$ must be a straight line because population growth is dependent on time." I now knew that I needed to back up and hit certain key ideas again.

Ferrini-Mundy and Graham [7] are among a number of researchers in mathematics education who have been examining student conceptions of such basic ideas as limits, derivatives, and integrals. Much of what they find can upset our notions of what we should be testing:

- Confusion between approximation as it is used in the definition of a limit and as it pertains to the actual limit: "It [$0.999\overline{9}$] wouldn't be quite 1 [but it is] close enough to solve the problems that we needed to solve."
- The ability to correctly locate local maxima, minima, and inflection points in order to sketch the graph of a function, while the same person has a complete inability to explain the relationship between the derivative and the tangent line.
- A recognition of the definite integral both as a process (a signal to anti-differentiate and substitute values) and as a representation of the area under a curve without having formed any connection between these two representations.

Their work can alert us to serious misconceptions that are common and which our current tests may not uncover. We have the responsibility to determine whether these misconceptions are present and, if so, to modify our own teaching accordingly.

Common Tools of Assessment

Traditional Tests, Quizzes, and Homework

If speed and accuracy in executing well-established algorithms are what we value most, then multiple-choice tests with less than five minutes per problem are appropriate. If we want to discover whether our students are capable of applying the ideas of calculus to unfamiliar problems, then we must give few problems and sufficient time to actually think about them. There is, of course, middle ground between these two extremes of what we want and what we feel we are forced to use by constraints of time or class size. The point is that we must be aware of how far what we test diverges from our true objectives so that we can continually improve the tests and so that we are aware of what other evaluative tools we will need if students are to perceive that we really do expect them to achieve the stated goals.

Student perception is very important. It is not enough that we can look over the questions we have written and see that they are consistent with our objectives. The students must also see this. This is the greatest flaw of the multiple-choice test. No matter how well the questions and possible answers have been constructed, most students believe that the test favors speed and accuracy above thoughtfulness and understanding. They also see the incorrect solutions as traps that we have set for those who make minor errors.[1] With this perception of our tests, it should not surprise us if students believe that the purpose of the calculus course is to reduce the number of science and engineering majors.

My examinations have very few but very challenging problems, and I allow students to improve their grade by correcting the answers that were wrong. I have also found it useful to supplement such examinations with a **gateway test**. This is a straightforward test of the techniques of differentiation or integration that students are allowed to retake as many times as necessary. The grade is not averaged into the course grade, but a pass at the level of 80% or 90% or 100% is required for passing the course. I require a pass at 100%, but I do test exactly the same set of techniques—with different problems—on each retest. Starting with the second retest, I let the student know which answers are incorrect and give the opportunity to change them before I grade the test. In two years of using gateway tests, every student has passed. I have also found that my weakest students appreciate this test the most; it forces them to a minimal level of competency. The experience of others is that those who never pass the gateway test are extremely rare. Usually, they are already failing the course. Occasionally, it may be expedient to drop the course grade by one letter rather than to fail them.

We should collect and grade **homework problems** if we expect them to be done faithfully and completely. If grading is not possible, then **quizzes** can be used as a check that the problems have been done. I use unannounced quizzes in addition to graded homework to

[1] These points were articulated in the survey of student attitudes [4].

verify that students have done the required reading and kept up with assignments. Such quizzes are short and do not address more than a single, simple idea.

Writing Assignments, Expositions, and Proofs

One of the best ways of evaluating student understanding is through the use of writing assignments. These can take a variety of forms, from a careful explanation of how a particular technique works and when it is appropriate to employ it to an extensive write-up of the solution to a demanding problem. It is not clear to me when we should begin to require the ability to read and analyze proofs, but whenever that time comes, students will understand the structure of a proof much faster if they write and rewrite their own proofs. Nothing forces the student to clarify a personal understanding and reveals where there are gaps in that understanding as effectively as the requirement to explain it in writing.

Very few students turn in mathematics assignments that are in any sense of the word readable. This does not necessarily imply an inability to write. I have known many students who turn in beautiful exposition when it is required, and a page of scrawl with circled answers when they feel that that is sufficient. We must be clear about our expectations, and we must be prepared to give the support that may be needed to achieve them. Most students have to learn how to write mathematics.

There are several things that aid in the development of the ability to write mathematics. One is to insist that students use the first person most of the time and active voice all of the time. Another is to critique early versions of the reports before they submit the draft that is to be graded, especially early in the year. It may be necessary to set aside time to talk about what makes for good writing in mathematics. And it helps if students are collaborating, at least on the first few assignments. They do learn from each other.

Major Projects and Laboratory Work

Projects give students the chance to exercise higher level thinking skills: to apply their knowledge, to integrate concepts, to synthesize new solutions, to evaluate the validity of the solutions they find. They can also make class exciting. Through them, I have seen students discover the creative and engaging side of mathematics. I find nothing in teaching so exhilarating as a classroom abuzz with small groups of students tackling a difficult problem.

Group work is the right way to handle most major projects. It builds confidence, it forces students to think through their insights as they explain them to their collaborators, it corrects many errors before they become embedded in the solution, and it facilitates discoveries that might not have been made by individuals. There are many ways of organizing such group work and of maintaining individual accountability. I leave the organization of each group up to the participants, shuffle the students after each project, and require them to sign the report, signifying that each has contributed to it. Once students are accustomed to writing such reports, it is useful to have some projects that are written up individually. This is most appreciated by the best students—who are given the chance to show what they are capable of doing alone—and the weakest students—who often feel that they have only been able to make minimal contributions to group projects.

The literature is now rich with such projects. The five volumes of *Resources for Calculus* [11] constitute one starting point.

Laboratory work is similar but has its own set of opportunities and hazards. At its best, it is strongly integrated into the classroom experience: a new concept is introduced in class, the next day students use computers or graphing calculators to explore it in various contexts and to begin to build an understanding of how it is used and how it relates to what they already know. They then return to the classroom where questions can be answered and the concept further developed. Student evaluations of the labs in *Project CALC* [2] indicated that this is what was happening:

> "Helpful in clearing up confusing concepts in text."

> "Several times I found myself thinking well, I'll just wait till Tuesday, maybe that will help, and it did."

But there are dangers in laboratory work. It usually has a fixed-time limit. It is difficult to make it challenging and engaging without putting it beyond what can be completed in the time allotted. To err in the other direction is worse. Labs lose their value if they are made mechanical, if they can be executed without forcing students to think about what is happening. And if they are not directly tied to the classroom—used and built upon by the classroom instructor—then they are meaningless.

Oral Presentations

It is not enough to be able to write well and to interact constructively in small groups. We would like our students to reach the level of familiarity and confidence at which they are able to present and defend solutions before an audience. This can be as simple as a homework problem whose solution is explained by one student, or as complex as the digestion of a special topic to be presented to the entire class. One way of guaranteeing everyone's participation in a group project is to hold each liable for presenting the group's solution to the rest of the class.

Class Journals and Individual Portfolios

Students take justifiable pride in their best work, but interest in what they have done quickly wanes once a grade has been assigned. One way of maintaining interest and pride is to publish a **class journal**, taking one or perhaps two examples of the very best reports from each major assignment. In addition to serving as recognition for students who have put in extra effort, it demonstrates to the entire class what you consider to be exemplary work.

Another means of preserving a student's best work is to require that each keep a **portfolio** of those reports or examinations that represent the student's best work. This enables subsequent instructors to see what the student has done, is capable of doing, and still needs to do. It helps the student track the progress that has been made.

Ungraded Assessment

One of the great myths of education is that the experienced college teacher can tell how well the class is following from the level of perceived attentiveness. That is simply not true. I have been teaching long enough to know how easily I can deceive myself by concentrating on the best students. We need other measures of what is happening in the classroom, not only for our benefit but to help our students recognize when they do not understand a concept that is important. Some of the techniques that I have found helpful are:

- Ask a question and force the students to come up with an answer that they are willing to defend. Take a few minutes for them to discuss their solution with neighbors. This works particularly well when you give a problem with two plausible answers and ask how many people favor one or the other of the alternatives. When first presented, usually very few will be sufficiently confident to raise their hands. After a few minutes of talking it over with neighbors, you will find that most of the class is now willing to state a preference (and that most have now correctly identified the right answer).
- In the last minute of the period, have everyone write down brief answers to two questions that ask for some sort of summary of the period: "What did we do this period?" and "What unanswered question do you still have?"
- Have a colleague sit in on your class and observe your students. If you really want information about student understanding, have your colleague interview selected students and probe what they believe they have learned in the course.

A good resource for assessment techniques is *Classroom Assessment Techniques* by Angelo and Cross [1]. It has a general overview of the topic of ungraded assessment together with fifty examples that each include a sample implementation, pros, and cons.

Ungraded assessment is almost always for our benefit. However modest or ambitious it may be, it is a waste of time unless we use the information obtained from it to improve our teaching.

Student Responsibilities

Having described all of the things that we should be doing, I want to make the point that it is *not* our responsibility to see that everyone achieves the goals we have set. This is where I have the greatest difficulty with the NCTM *Assessment Standards for School Mathematics* [10] which cites "equity" as one of the standards. Assessment in the classroom does have a role to play as filter, identifying those students who have fallen short of the goals. Our goals must be constructed with care: they must be honest, they must be realistic, they must be challenging. And they must be upheld. A large part of the purpose of assessment is to hold each student accountable for making satisfactory progress toward these goals.

If we insist on this accountability, we can go a long way toward correcting one of the great problems in our calculus classes: the meager amount of time that most students spend studying calculus. From my own survey of students at The Pennsylvania State University [4], I have seen that the problem is more one of knowing what to do when studying than a lack of willingness to study. Our students come to college with the belief that studying mathematics means doing the exercises at the end of the section. Furthermore, experience has taught them that the really hard problems seldom appear on tests. I was surprised to learn how very conscientiously almost all students will do the relevant problems the same day they have been discussed in class, even when these exercises will not be collected or graded. But that is all they know to do. The surveyed students spend about an hour working these assigned problems. When they start to go beyond the examples worked out in class—and thus beyond what they believe will be on the test—most students will stop working. They have other demands on their time. If the only assessment tool we use is the traditional test, then the only studying most students will do is what has worked for them on tests in the past.

A common complaint from students in nontraditional calculus classes is that too much work is required. Limited data suggests that the C4L program of Dubinsky, Schwingendorf, and Mathews comes close to doubling the amount of time students spend studying calculus [8]. As we broaden student perception of what they are required to know and be able to do in mathematics, we can expect that they will study more and get more out of the time that they spend studying.

Student responsibilities are real and cannot be ignored, but we must first show them how to study effectively and then hold them accountable for so doing.

Conclusion

What I have described in this paper is at the opposite pole from the ideal with which I began. My classes will usually involve tests, graded homework assignments, unannounced quizzes, gateway tests, major projects, lab reports, written exposition and proofs, as well as various forms of ungraded assessment. It begins to look like a class that is so busy getting educated that no one has time to learn. But my own experience has been that something amazing emerges when evaluation is this constant and this varied: *the evaluation ceases to be the focus of the course.*

Students no longer study just for the test. Because they are held accountable for all aspects of the course, they work at all aspects of the course. Because no single evaluation will determine their grade, no single evaluation dominates their horizon. They learn what it means to understand mathematics, and they begin to deepen their own understanding.

Teaching in this way is unsettling, and it means more work for the professor. It is uncomfortable to learn how ineffective our beautiful lectures can be. It takes time to determine where and how we can improve what we are doing and to settle into new modes of teaching. The increased assessment is time consuming. There are ways to do the assessment efficiently: collaborative projects rather than individual ones reduce the number of papers, some assignments can be self-graded or critiqued by other students, hired graders can be used intelligently, but it still requires more effort on our part than the traditional lecture-exam format. The question is not whether we can afford this extra effort, it is whether we can afford not to take assessment this seriously.

References

[1] Thomas A. Angelo & K. Patricia Cross, *Classroom Assessment Techniques: a handbook for college teachers*, 2nd edition, Jossey-Bass, San Francisco, 1993.

[2] William H. Barker, *Mathematica Laboratory Manual to accompany The Calculus Reader*, revised preliminary edition, D. C. Heath, 1994.

[3] B. S. Bloom *et al.*, *Taxonomy of Educational Objectives*, vol. 1: *Cognitive Domain*, McKay, New York, 1956.

[4] David Bressoud, Student Attitudes in Calculus, *Focus*, **14**, June 1994, 6–7.

[5] Committee on the Undergraduate Program in Mathematics, *Assessment of student learning for improving the undergraduate major in mathematics*, The Mathematical Association of America, Washington, DC, 1995.

[6] John Ewing, Comment, *The American Mathematical Monthly*, **102**, Feb. 1995, 98.

[7] Joan Ferrini-Mundy & Karen Graham, Research in calculus learning: understanding of limits, derivatives, and integrals, *Research Issues in Undergraduate Mathematics Learning*, Kaput & Dubinsky, editors, MAA Notes #33, The Mathematical Association of America, Washington, DC, 1994.

[8] David Mathews, Time to study: the C4L experience, to appear in *UME Trends*, 1995.

[9] Lawrence C. Moore & David A. Smith, *Project CALC Instructor's Guide*, preliminary edition, D.C. Heath, 1992.

[10] National Council of Teachers of Mathematics, *Assessment Standards for School Mathematics*, working draft, Reston, VA, 1993.

[11] *Resources for Calculus*, vols. 1–5, A. Wayne Roberts, Project Director, MAA Notes numbers 27–31, The Mathematical Association of America, Washington, DC, 1993.

[12] Patricia M. Schwarz, Feynman's Theorem on Being a Student, INTERNET newsgroup SCI.MATH, 26 Dec. 1993.

Final Examination for Calculus 1: Test 1

You have until noon to complete this exam. Work neatly on separate paper and explain your work.

1. The graph of $y = f'(x)$ is given below.

   (a) On what intervals is f increasing? decreasing? Why?

   (b) On what intervals is f concave up? concave down? Why?

   (c) Assuming that f goes through the point $(0, 2)$, sketch a graph of f on the axes below.

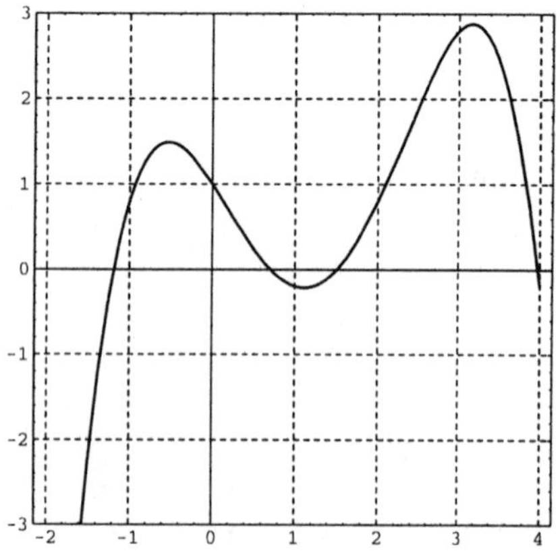

2. Consider the region **R** bounded by $y = e^x$, $y = e^3$, and $x = 2$.

   (a) Sketch the boundary curves and shade in **R**.

   (b) Find the area of **R**.

   (c) Find the average height of **R**.

   (d) Set up an integral for the volume of the solid obtained by rotating **R** around the x-axis. Do not integrate.

3. Let $f(x) = \sqrt[3]{x^2 - 1}$.

   (a) Write the equation of the tangent line to this curve at $x = 3$.

   (b) We discussed in class that this tangent line is a good approximation to the function for values of x close to 3. Use your answer to part (a) to approximate $\sqrt[3]{8.61} = f(3.1)$.

4. The graph of a function h is shown below.

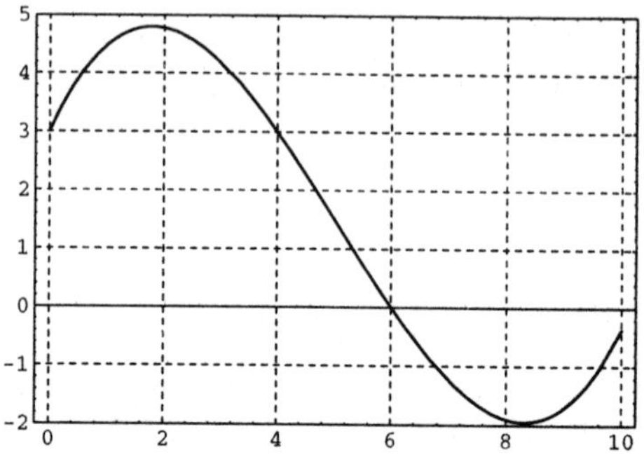

   List, from smallest to largest:

   (i) the average value of h over the interval $[0, 10]$;

   (ii) the average rate of change of h over the interval $[0, 10]$;

   (iii) $h'(5)$;

   (iv) $\int_0^{10} h(x)\,dx$;

   (v) $\int_0^{5} h(x)\,dx$;

   (vi) $\int_6^{10} h(x)\,dx$.

   Explain your reasoning.

5. Suppose that the materials for a picture frame cost the manufacturer \$0.45 per inch, except that the stronger materials for the top side cost \$0.75 per inch. Find the dimensions of the cheapest frame that includes an area of 120 in^2.

6. A bacteria culture is known to grow at a rate proportional to the number present. At 1 p.m., 1000 bacteria are present. At 4 p.m., 3000 are present.

   (a) Find the number of bacteria present at noon. (Hint: Let $t = 0$ correspond to 1 p.m.)

   (b) Find the average number of bacteria present between 1 p.m. and 4 p.m.

7. Short answer problems.

(a) $\int_0^1 e^x\sqrt{e^x+1}\,dx$

(b) $\int \frac{2x+3}{x^2+3x+7}\,dx$

(c) Verify $\int x^2e^x\,dx = x^2e^x - 2xe^x + 2e^x + C$

(d) Verify $\int \frac{1}{9-x^2}\,dx = \frac{1}{6}\left(\ln|x+3| - \ln|x-3|\right) + C$

8. A pyramid 100 feet tall has a square base. The perimeter at height y is $(100 - y)$ feet. (For example, the perimeter of the square at the bottom is 100 ft. The perimeter of the square 20 feet from the bottom is 80 ft.)

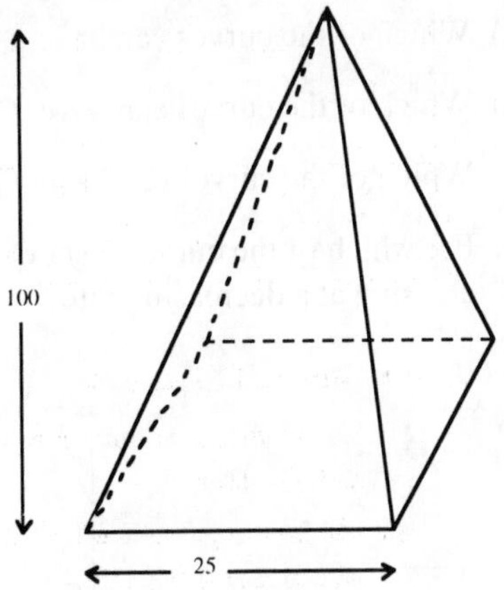

(a) Approximate the volume of the pyramid using 5 equally spaced measurements. In other words, approximate the pyramid by 5 rectangular blocks stacked upon one another. Your blocks should all have square bases, which decrease as one goes up. The blocks should all have the same height.

(b) Write a Riemann sum to approximate the volume of the pyramid using n equally spaced measurements.

(c) Based on part (b), write an integral that represents the volume of the pyramid.

(d) Evaluate your integral from part (c) and thus find the exact volume of the pyramid.

9. The temperature of an object at time t is given by

$$T(t) = 0.1\left(t^4 - 12t^3 + 2000\right), \quad 0 \le t \le 10.$$

(a) Find the hottest and coldest temperature during the interval [0, 10].

(b) At what time is the rate of change in the temperature a minimum?

10. Suppose the graph of f over [0, 2] is given in the figure below.

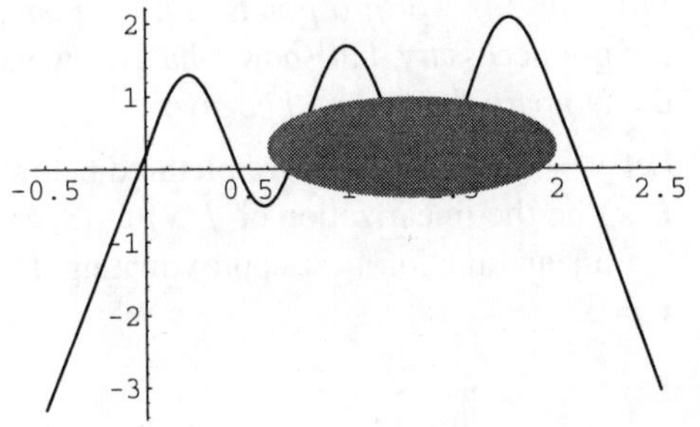

As you see, a section of the graph is concealed. Decide whether each of the following definite integrals is positive, equal to 0, or negative. Explain your reasoning.

(a) $\int_0^2 f(x)\,dx$ (b) $\int_0^2 f'(x)\,dx$

Final Examination for Calculus 1: Test 2

1. *Each part of problem 1 is worth 5 points. Full explanations are* not *necessary, but show whatever work you do. No partial credit will be given.*

   (a) Find the equation of the line tangent to $f(x) = x^3 + 1$ at $x = 2$.

   (b) Find dy/dx for the equation $xy + y^2 = 3$.

   (c) Calculate $\int_{0.25}^{0.45} (2^x + 4)\, dx$. Use 50 subdivisions.

   (d) If $y = 1/x$, find d^2y/dx^2.

2. *Problem 2 is worth 6 points. Full explanations are* not *necessary, but show whatever work you do. No partial credit will be given.*

   Let $y = f(x)$ be the curve pictured below. Let $L(x)$ be the linearization of $f(x)$ at $(3, 4)$, i.e., the linear function best approximating $f(x)$ at $x = 3$.

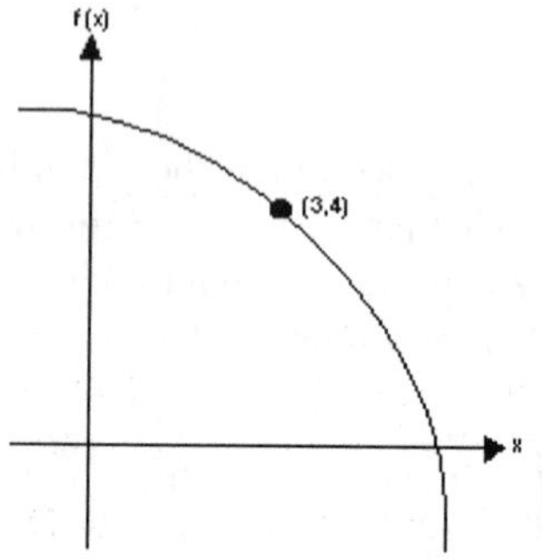

   Circle the correct answer:

   (a) The slope of $f(x)$ at $x = 3$ is the slope of $L(x)$ at $x = 3$.

   True
   False
   Not enough information given

   (b) $L(x) = 4(x - 3) + 4$

   True
   False
   Not enough information given

3. *Problem 3 is worth 12 points. Full explanations are* not *necessary, but show whatever work you do. No partial credit will be given.*

   Consider the curves below, labeled A–H. For each of the following problems, list all the curves that apply. Note that for each curve, the scales on the x-axis and the y-axis do not have to be the same.

   (a) Which of the curves can have $f'(P) \geq 0$?

   (b) Which of the curves can have $f''(P) = 3$?

   (c) Which of the curves can have $f''(P) = 0$?

   (d) For which of the curves is $f(x)$ always increasing at a decreasing rate?

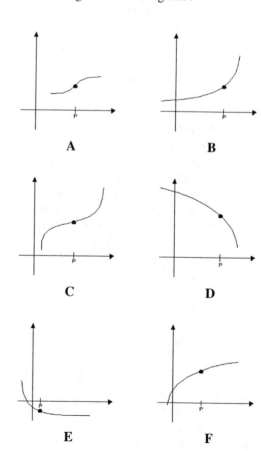

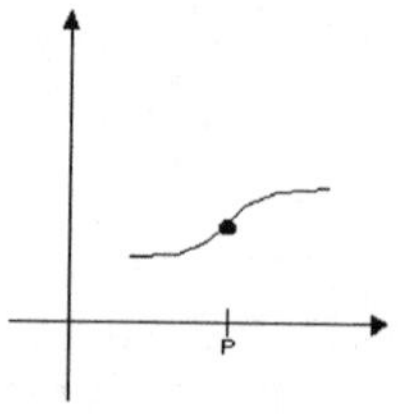

A

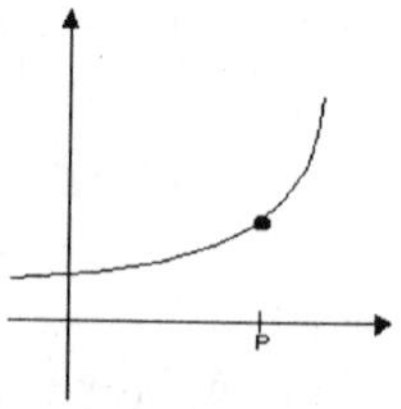

B

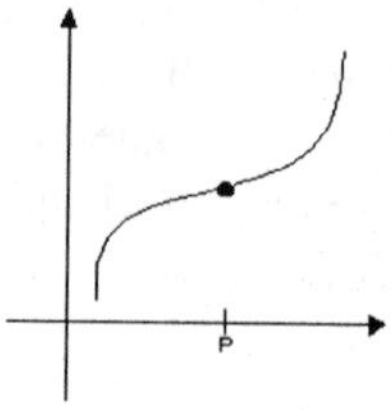

C

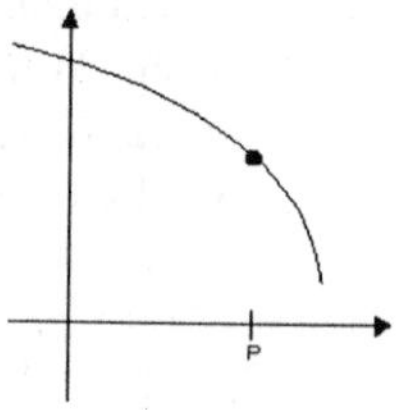

D

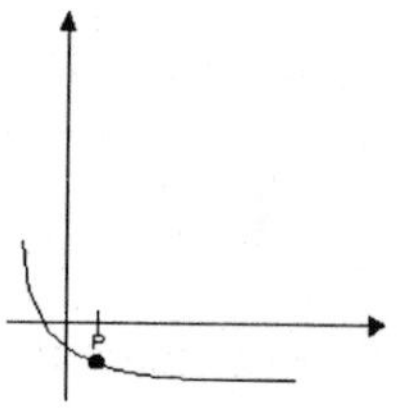

E

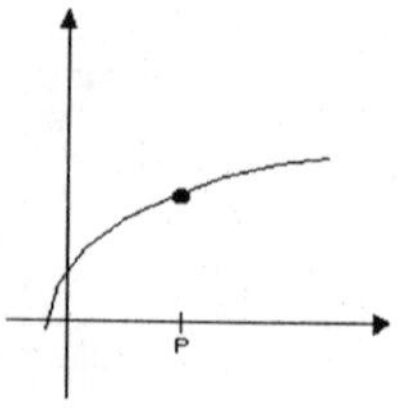

F

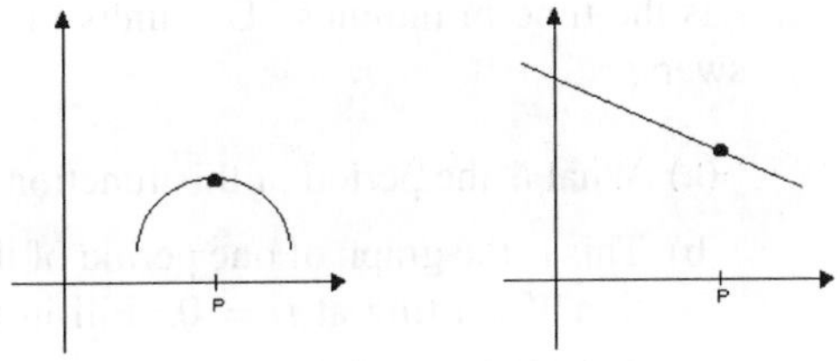

G **H**

4. *Problem 4 is worth 5 points. Full explanations are not necessary, but show whatever work you do. No partial credit will be given.*

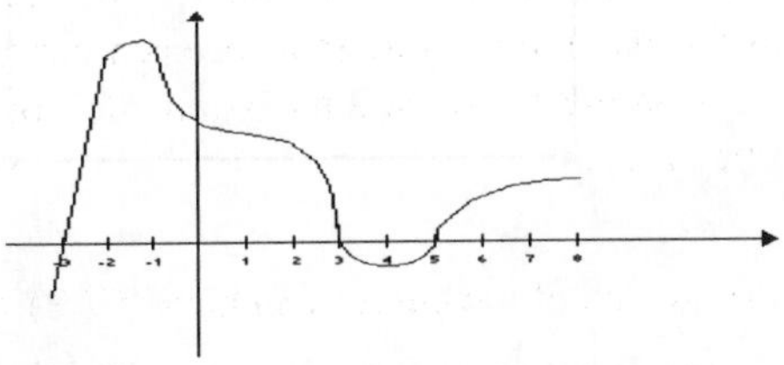

The graph above is of the function $y = f(x)$. Put the following in increasing order:

$$A = \int_{-3}^{3} f(x)\,dx \qquad C = \int_{3}^{5} f(x)\,dx$$

$$B = \int_{0}^{3} f(x)\,dx \qquad D = \int_{0}^{5} f(x)\,dx$$

_____ < _____ < _____ < _____

5. *Problem 5 is worth 8 points. Full explanations are not necessary, but show whatever work you do. Limited partial credit may be given.*

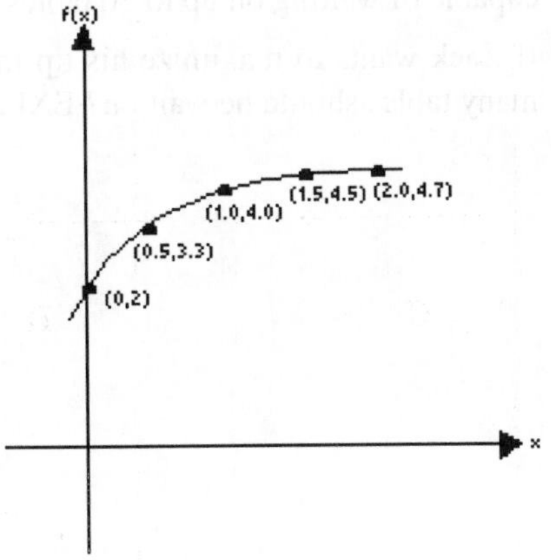

Consider the graph above. Write the following Riemann sums for the integral $\int_0^2 f(x)\,dx$. Use four intervals.

(a) Right hand sum.

(b) Left hand sum.

(c) Use your answers from parts (a) and (b) to give your best estimate for $\int_0^2 f(x)\,dx$.

6. *Problem 6 is worth 10 points. Make sure we understand how you arrived at your answers. Limited partial credit may be given.*

t	$f(t)$
8	8.253
9	8.372
10	8.459
11	8.616

(a) Estimate $f'(10)$.

(b) Estimate $f^{-1}(8.5)$.

(c) Evaluate $\int_8^{10} f'(t)\,dt$.

7. *Problem 7 is worth 7 points. Make sure we understand how you arrived at your answers. Limited partial credit may be given.*

There's dust on my guitar! The total amount of dust after t days is given by $g(t)$. I know that $g(30) = 270$ milligrams and $g'(30) = 5$.

(a) Estimate $g(32)$.

(b) What are the units of $g'(t)$?

8. *Problem 8 is worth 12 points. Make sure we understand how you arrived at your answers. Limited partial credit may be given.*

Water is draining from a tank. The volume of water in the tank is given by

$$V(t) = 1000 + (20 - t)^3,$$

where V is in gallons and t is the number of hours since the water began draining. Make sure your answer contains the correct units.

(a) How much water is in the tank initially?

(b) How fast is it draining after 10 hours?

(c) Will the tank have been completely drained after two days? Why?

9. *Problem 9 is worth 15 points. Full explanations are not necessary, but show whatever work you do. Limited partial credit may be given.*

   An experimental jet car runs along a track for 6 seconds before exploding into a giant fireball that could be seen for miles. From its starting time, $t = 0$ seconds, its speed in feet per second is given by the formula $V(t) = 0.08te^t$. Make sure your answers contain the correct units.

   (a) How fast was it going at $t = 6$ seconds?

   (b) What was its average acceleration over the first 6 seconds?

   (c) How far did it travel during the first 6 seconds?

10. *Problem 10 is worth 10 points. Make sure we understand how you arrived at your answers. Limited partial credit may be given.*

    You are given $F(x) = x^2 \ln x$.

    (a) You overhear a laboratory mouse saying, "It is clear that $F'(x) = x(1+2\ln x)$." Verify that he is right.

    (b) Use part (a) and the Fundamental Theorem of Calculus to calculate $\int_1^2 x(1+2\ln x)\,dx$.

11. *Problem 11 is worth 10 points. Make sure we understand how you arrived at your answers. Limited partial credit may be given.*

    When hyperventilating, a person breathes in and out extremely rapidly. A spirogram is a machine that draws a graph of the volume of air in a person's lungs as a function of time. During hyperventilation, the spirogram trace might be represented by $V(t) = 3 - 0.05\cos(200\pi t)$, where V is the volume of the lungs in liters and t is the time in minutes. Use units in your answers.

    (a) What is the period of this function?

    (b) This is the graph of one period of the function V starting at $t = 0$. Fill in the four missing coordinates.

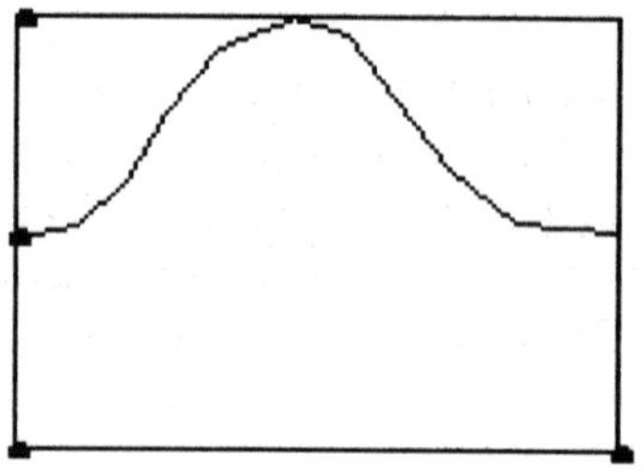

    (c) What are the maximum and minimum volumes of air in the lungs?

    (d) Find the maximum rate (in liters/minute) of flow of air during inspiration (i.e., breathing in).

12. *Problem 12 is worth 10 points. Please give a full explanation. Partial credit will be given.*

    Zack is a waiter at a restaurant with 100 tables. During his first month he waited on 20 tables every night, and collected an average tip of \$15 from each table. He started to work more tables, and noticed that for every extra table he took on in a night, his average tip would go down 25 cents per table. He figures that he is physically capable of waiting on up to 30 tables in a night.

    If Zack wants to maximize his tip money, how many tables should he wait on? EXPLAIN.

Final Examination for Calculus 1: Test 3

Part I

You have 45 minutes for this part of the exam, but you may go on to Part II as soon as you are finished with Part I. No calculators are allowed until Part I has been turned in; no books or notes are allowed during the entire exam. There are 52 points in Part I and 98 points in Part II.

1. Find formulas for the derivatives of the following functions. Answers should read *derivative = formula* where *derivative* is written with the appropriate f' or Leibniz notation.

 (a) $F(u) = \pi u^3 + \dfrac{1}{4u} + 3^u + \sqrt{7}$

 (b) $x = t^4 \cos(5t)$

 (c) $y = \dfrac{(\ln x)^6}{\tan x}$

 (d) $w = \ln \sqrt{\dfrac{z(z+1)}{z-3}}$

 (e) $g(x) = e^{\sqrt{x}}$

 (f) Suppose $z = f(x, y) = x^{3/5} \sin(xy^2)$. Compute $f_x = \partial z/\partial x$ and $f_y = \partial z/\partial y$.

2. Sketch the graphs and indicate the intercepts. Simplify the expression first, if necessary.

 (a) $y = \ln \dfrac{1}{x}$

 (b) $y = e^{-2x}$

 (c) $y = \ln(2e^{0.5x})$

Part II

On this part of the exam no notes or books are allowed, but calculators are OK. Be sure to show your work for full credit.

1. Find a formula for $F(t)$ if $F'(t) = 8t^7 - 6t^2$ and $F(1) = -7$.

2. (a) Write the microscope equation for $y = 27/(x+1)$ at $x = 2$.

 (b) Write the equation for the tangent line to the graph of $y = 27/(x+1)$ at $x = 2$.

3. Suppose $F(1) = 2$ and $0 \le F'(x) \le 1$ for $0 \le x \le 3$. Sketch and shade the smallest region in which you can guarantee that the graph $y = F(x)$ must lie.

4. You drop a pebble off a bridge into a stream. Beginning with an initial velocity of 0, it falls faster and faster under the pull of gravity, but when it enters the water it slows down to a constant terminal velocity, which it maintains until it hits bottom. Sketch a graph of the pebble's distance from your hand as a function of time up to the time of impact.

5. Suppose f is locally linear at $x = -2$, $f(-2) = 4$ and $f'(-2) = -3$. Estimate $f(-1.7)$.

6. For each function whose graph $y = f(x)$ is sketched below, sketch the graph of the derivative function $y = f'(x)$ on the same set of axes.

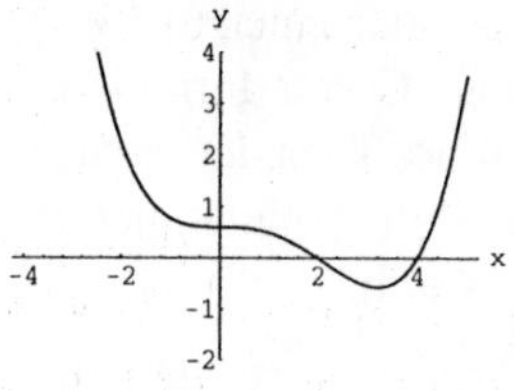

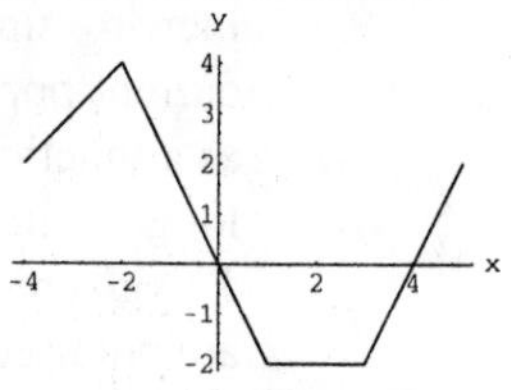

7. Measurements of the length L (in cm) of a rod are made at different temperatures T (in °F), as shown in the table.

T	56	58	60	62	64
L	8.795	8.902	9.011	9.121	9.235

 (a) Using at least four values from the table, compute estimates of the rate of change of L with respect to T when $T = 60°$F. Based on these estimates, what is $L'(60)$ to three decimal places?

 (b) Estimate the temperature at which the rod is 9 cm long.

8. Verify that $y = (C + 6x)^{1/3}$ satisfies the DE $y' = 2/y^2$, and find the value of C so that $y(1) = 3$.

9. If f is differentiable at every real number x, $g(x) = x^2 f(x)$, $f(1) = 3$ and $f'(1) = -2$, compute $g'(x)$ and then $g'(1)$.

10. The population of a variety of truffle increases at an annual rate equal to 3% of the current population P (which is measured in kilograms per acre); this takes into account the birth rate and the death rate due to natural causes. Meanwhile humans harvest 0.5 kg/acre each year.

    (a) Write a differential equation that describes the rate of change of the truffle population with respect to time, taking both the natural increase and the loss to human consumption into account.

    (b) If the current population is 10 kg/acre, use Euler's method with two steps to approximate the population four years from now.

    (c) Scarcity of the truffles leads to a call for halting the harvesting until the population reaches 12 kg/acre. What amount should then be harvested each year to exactly maintain this population?

    (d) Assume that the population is 10 kg/acre and there is no harvesting, so the truffles increase simply at a rate equal to 3% of the current population. Give a formula for P as a function of time. Then determine how long it will be until the population reaches 12 kg/acre. (Parts a, b, c, of this problem are not used here.)

11. A basin in the shape of a parabolic bowl is used as a reservoir. The volume is $V = \frac{1}{2}r^2h$ if r is the radius of the surface and h is the depth.

    (a) Compute the partial derivatives of V with respect to r and h and then write the total microscope equation for ΔV.

    (b) Express the relative error in measurement of V in terms of the relative error in measurement of r and h.

    (c) If silt begins to cover the bottom, so h decreases by 6%, what change in r will have to be made to compensate if we want the volume to remain the same?

12. A truck driver starts a trip by driving at 50 mph for 2 hours and then cruises along at 60 mph for 1.5 hours. She then stops for lunch, which takes a half hour, and finishes the trip by driving at 30 mph for another hour. (Assume the changes in speed take place instantly.)

    (a) Graph the speed of the truck as a function of the time t in hours.

    (b) Graph the distance traveled D as a function of time t.

    (c) Find a formula for the distance D as a function of t for the last hour.

    (d) What is the average speed for the entire trip? It appears that our driver was never going this speed. What small change in our description of the journey would allow us to conclude that she did hit this average speed at least once?

13. Consumer acceptance of a new VCR depends on the price; in fact $N = e^{-2p}$ is the fraction of potential customers who will purchase a unit if it costs p dollars.

    (a) If the price is \$0, what is N? Does this make sense? Use a graph to describe what happens to N as the price p increases.

    (b) On average, each customer brings in revenue of $R(p) = pN = pe^{-2p} = p/e^{2p}$ dollars. The most "sales" are made if $p = 0$, but obviously this does not produce the maximum revenue. What happens to R if p is very large?

    (c) Compute $R'(p)$ and determine for which values of p the revenue is increasing.

    (d) Sketch R as a function of p. At what price level does one obtain the maximum revenue?

Final Examination for Calculus 1: Test 4

Take Home Portion

1. Let $f(x) = x^{\sin x}$.

   (a) Using your calculator, estimate $f'(2)$ and explain your method. (Be sure you are in a radian mode!)

   (b) Use your result from part (a) to get a linear approximation for $(2.1)^{\sin 2.1}$.

   (c) Is your estimate in (b) greater or smaller than the actual value? Explain.

2. If P dollars are invested at an annual interest rate of $r\%$, then in t years this investment grows to F dollars, where $F = P(1 + r/100)^t$.

   (a) Find dP/dt by solving the equation for P and differentiating assuming F and r to be constants.

   (b) What is the sign of dP/dt? Explain why this is reasonable.

3. (a) For $f(x) = x^2e^{-x}$ find $f'(x)$ and determine all local maxima and minima.

   (b) Determine global (absolute) maxima and minima for f.

   (c) Determine the "end behavior" of the function and sketch a graph of f which incorporates all of your results and label key points. You should be able to do this without your calculator.

4. A Math 105 (Pre-calculus) student asks you to explain the concept of the derivative. In vocabulary appropriate to the situation and in your own words, give a brief, concise explanation to the student.

5. One fine day you take a hike up a mountain path. Using your trusty map you have determined that the path is approximately in the shape of the curve: $y = 4x^3 - 48x^2 + 192x + 144$. Here y is the elevation in feet above sea level and x is the horizontal distance in miles you have traveled, but **your map only shows the path for 7 miles**, horizontal distance.

   (a) How high above sea level do you start your hike?

   (b) How high above sea level are you at the beginning of the 7th mile?

   (c) Use your calculator to draw an **informative** graph of the path and sketch your answer below. Show the scale you use.

   (d) Without the calculator, find where on the path is a nice flat spot to stop for a picnic. Explain.

   (e) Estimate the elevation after 7.5 miles. Reminder: You do not know the shape of the path beyond 7 miles!

   (f) Your friend, who is *not* in "good shape," followed the path for 15 miles total horizontal distance the previous day. Does it make sense for the equation for the elevation to continue to hold much beyond the 7 mile mark? Explain.

6. During Professor E's days as a student last century, he often studied his calculus in a dim, unheated room with a single lit candle. One particular day in mid-winter, after walking 10 miles uphill (both ways, of course!) through knee-deep snow to attend class, he returned home too tired to study. After lighting the solitary candle on his desk, he walked directly away cursing his woeful situation. The temperature (in degrees Fahrenheit) and illumination (in % of candle power) decreased as his distance (in feet) from his desk (and candle) increased. In fact, I kept a record and have displayed the information in the table below, just in case you may not believe the preceding sad tale!

Distance (feet)	Temperature (°F)	Illumination (% candlepower)
0	55.0	100
1	54.5	88
2	53.5	77
3	52.0	68
4	50.0	60
5	47.0	56
6	43.5	53

I get cold when the temperaure is below 40°F and it is dark when the illumination is at most 50% of one candlepower.

(a) Which graph below is the graph of temperature as a function of distance and which is the illumination as a function of distance graph?

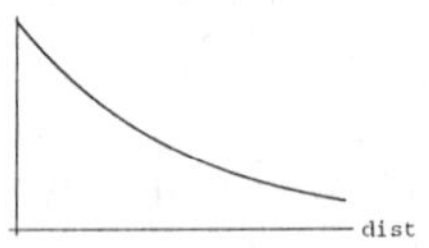

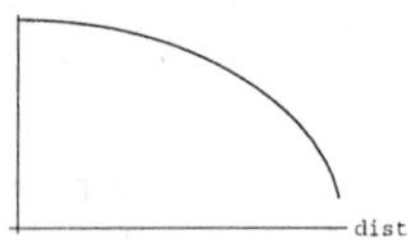

(b) What is the average rate at which the temperature is changing when the illumination drops from 77% to 56%?

(c) I can still read my watch when the illumination is 6.4%. Can I still read my watch when I am 3.5 feet from the solitary candle? Explain.

(d) Suppose that at 6 feet the instantaneous rate of change of the temperature is -4.5°F/ft and the instantaneous rate of change of the illumination is -3% candle power/ft. Estimate the temperature and the illumination at 7 feet.

(e) Am I in the dark (I know what you are thinking!) before I am cold or vice versa? Be sure and explain your answer.

In-Class Portion

1.

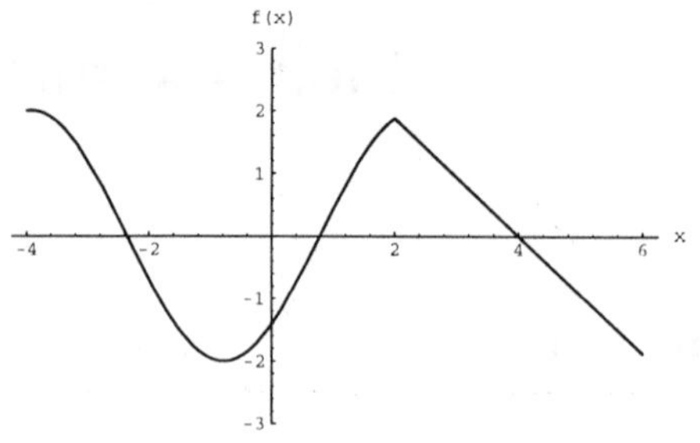

(a) Approximate the zeros of the function from the graph above.

(b) What is $f(-2)$ approximately?

(c) Is the function increasing or decreasing at $x = 3$?

(d) Is the graph concave up or concave down at $x = -4$?

(e) What is $f'(4)$ approximately?

(f) Are there any values of x where $f(x)$ is not differentiable? Explain.

2. (a) On the first axes below sketch a curve whose slope is both positive and decreasing at first, but later on the slope is both positive and increasing.

(b) On the second axes graph the first derivative of the curve in (a).

(c) On the third axes graph the derivative of $f'(x)$.

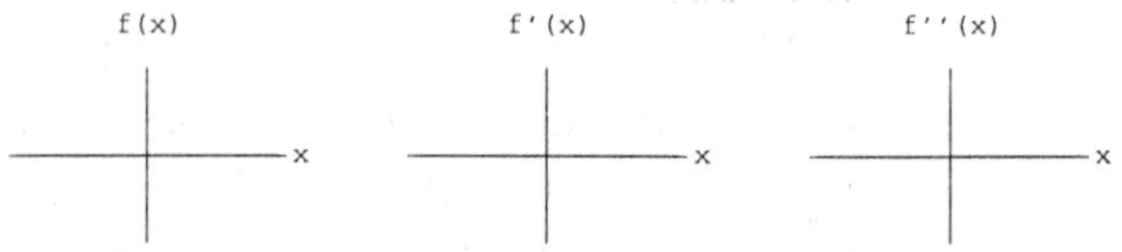

3. Find the derivative of the following functions. Show your work.

(a) $f(x) = \sqrt[4]{x^3} - \dfrac{2}{x} - 3\sqrt{x}$

(b) $t = \sqrt{w^2 + 5}$

(c) $f(\theta) = \sin(2\theta^3 + 1)$

(d) $r = \dfrac{e^{z^2}}{z^2 - 3}$

(e) $y = (x + \sin x)^{\pi}$

(f) $f(x) = 2^x \ln x$

4.

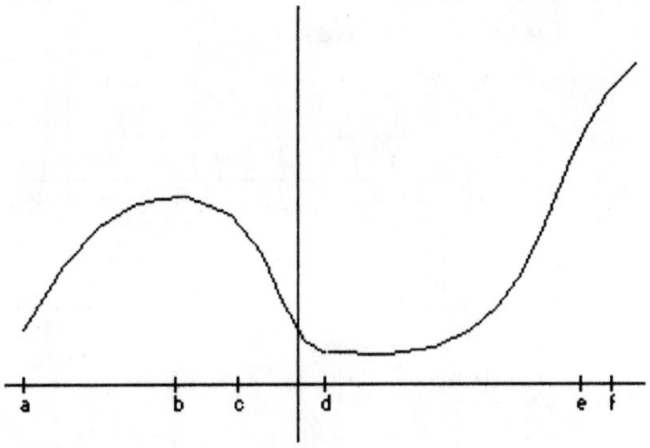

Above is the graph of $f'(x)$, **not** $f(x)$! At which of the marked values is:

(a) $f(x)$ greatest
(b) $f(x)$ least
(c) $f'(x)$ greatest
(d) $f'(x)$ least
(e) $f''(x)$ greatest
(f) $f''(x)$ least

Note: $f''(x)$ is the derivative of $f'(x)$.

5. Water is pouring into each of the jars in the top row of the diagram below at the constant rate of 5 gallons per minute. The graphs in the second row are the graphs of the **heights** of the water as a function of time. Write the letter below each graph inside its corresponding jar.

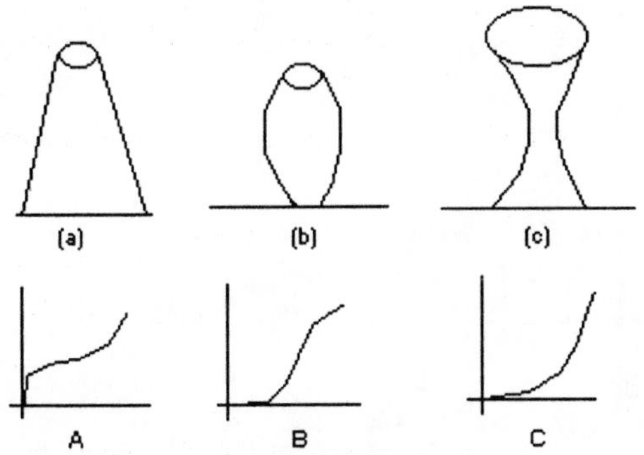

6. The yield from an apple orchard during a dry summer ($\approx$ 90 days) is dependent on the number of days the orchard is irrigated. Let $Y(x)$ be the yield in bushels of apples per tree when the orchard is irrigated x days during the summer.

(a) Interpret $Y(30) = 15$.

(b) Interpret $Y'(30) = 0.5$.

(c) Interpret $\frac{dY}{dx}\,|_{x=85} = -1$.

7. A particle moves in such a way that $x(t) = 4t^2 + 7\sin t$.

(a) What is the instantaneous rate of change at $t = 0$?

(b) What is the instantaneous rate of change at $t = \pi/2$?

(c) What is the average rate of change between $t = 0$ and $t = \pi/2$?

8. Suppose a function is given by the table of values below.

x	1.1	1.3	1.5	1.7	1.9	2.1
$f(x)$	12	15	21	23	24	25

(a) Estimate $f'(1.7)$.

(b) Write an equation of the tangent line to f at $x = 1.7$.

(c) Use your results above to estimate $f(1.8)$.

9. Solve $\ln x = \frac{1}{x}$ using your calculator and Newton's method.

(a) Complete the table below with your successive approximations (4 decimal places).

n	0	1	2	3
x_n	1.5000			

(b) If you found x_4, how accurate would it be? Explain.

10. The graph of the **derivative** of a function f is shown below.

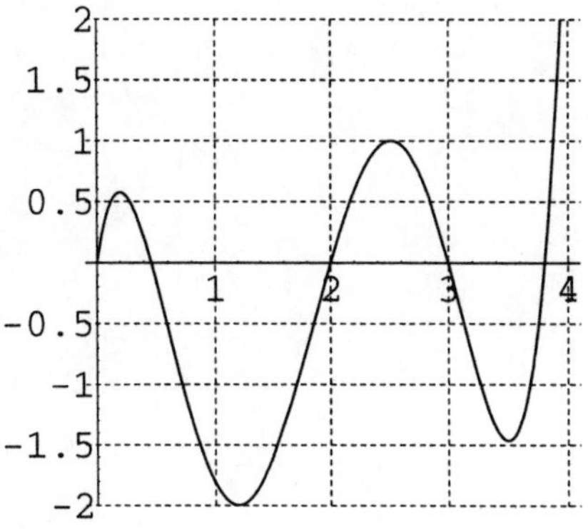

(a) State the intervals on which f is increasing and on which it is decreasing.

(b) Say where the local extrema occur, and for each one say whether it is a local maximum or a local minimum.

(c) EXTRA: Where in the interval $0 \le x \le 4$ does f achieve its global maximum?

(d) EXTRA: Suppose you are told $f(0) = 1$. Estimate $f(2)$.

Do 2 of the following 3 problems. You may do the other for extra credit.

1. Climbing health care costs have become a real source of concern. The data below shows the average yearly per capita health care expenditures for 1980 and 1985.

Year	Per Capita Expenditures ($)
1980	1055
1985	1596

(a) Use this data to estimate the average per capita expenditures in 1988 assuming health care costs are growing linearly.

(b) Use this data to estimate the average per capita expenditures in 1988 assuming health care costs are growing exponentially.

2. The table gives values for functions f and g and for their derivatives.

x	-1	0	1	2	3
f	3	3	1	0	1
g	1	2	2.5	3	4
f'	-3	-2	-1.5	-1	1
g'	2	3	2	2.5	3

(a) Find $\dfrac{d}{dx}\big(f(x)g(x)\big)$ at $x = 1$.

(b) Find $\dfrac{d}{dx}\Big(g\big(f(x)\big)\Big)$ at $x = 0$.

3. Suppose that $f(T)$ is the cost to heat my house, in dollars per day, when the outside temperature is T degrees.

(a) What does $f'(23) = -0.17$ mean?

(b) If $f(23) = 7.54$ and $f'(23) = -0.17$, approximately what is the cost to heat my house when the outside temperature is $20°$?

Final Examination for Calculus 1: Test 5

Instructions: This exam has three parts: Calculations, Concepts, and Essay Questions. We suggest the following time allocations (which also reflect the relative importance of each section in our grading): Calculations, 30 minutes; Concepts, one hour; Essay, one hour. If you have time remaining, use it for checking, refinement, and correction.

If you find that a calculation leads to an unreasonable answer, you will get more credit for identifying the apparent problem and saying why the answer is unreasonable, less credit for not noticing or for pretending there is no problem. Of course, you will get still more credit if you find what went wrong in the calculation and fix it. You may use your text, project, and lab materials, your tables, your calculator, and your notes.

Part I: Calculations

Show your work.

1. Find the derivatives of the following functions:
   (a) $f(x) = 3e^{-4x}$
   (b) $f(x) = \sin(x^2)$
   (c) $f(x) = \dfrac{x^2}{1+2x}$
2. Find (approximately) a number x which solves the equation $x = 5\ln x$.

Part II: Concepts

Write your answers in complete, connected sentences. (Note: Individual instructors selected problems for this section from among the following. No one instructor used all of these problems.)

1. At a certain instant, just before lifting off, a plane is traveling down the runway at 285 kilometers per hour. The pilot suddenly realizes something is wrong and aborts the takeoff by cutting power and applying the brakes. Assume that the effect of this action is a deceleration proportional to time t. After 28 seconds the plane comes to a stop. How far does the plane travel after the takeoff is aborted?
2. A patient's "reaction" $R(x)$ to a drug dose of size x is given by a formula of the form $R(x) = Ax^2(B - x)$, where A and B are positive constants. The "sensitivity" of the patient's body to a dose of size x is defined to be $R'(x)$.
   (a) What do you think the domain of x is? What is the physical meaning of the constant B? What is the physical meaning of the constant A?
   (b) For what value of x is R a maximum?
   (c) What is the maximum value of R?
   (d) For what value of x is the sensitivity a maximum?
   (e) Why is it called "sensitivity?"
3. A truck traveling on a flat interstate highway at a constant rate of 50 mph gets 4 miles to the gallon. Fuel costs \$1.15 per gallon. For each mile per hour increase in speed, the truck loses a tenth of a mile per gallon in its mileage. Drivers get \$27.50 per hour in wages, and fixed costs for running the truck amount to \$12.33 per hour. What constant speed should a dispatcher require on a straight run through 260 miles of Kansas interstate to minimize the total cost of operating the truck?
4. For what point(s) on the curve $y = x^2$ do(es) the tangent line to $y = x^2$ go through the point $(3, 5)$.
5. A spring of mass 4 kilograms and spring constant 9 kg/sec^2 is pulled in the direction away from a wall one meter beyond the equilibrium point. It is released from this point with an initial velocity (towards the wall) of 1.5 meters/sec.
   (a) Write down the differential equation and initial conditions satisfied by the function, $x(t)$, which gives the displacement from equilibrium as a function of time.
   (b) Solve to find $x(t)$.
   (c) At what times will the mass be closest to the wall?

Part III: Essay Questions

Answer each question in the form of a short essay.

1. Below are three graphs. One is the graph of $f(x)$, another is $f'(x)$, and another is $f''(x)$. Identify which is which and carefully explain your reasons in a paragraph or two.

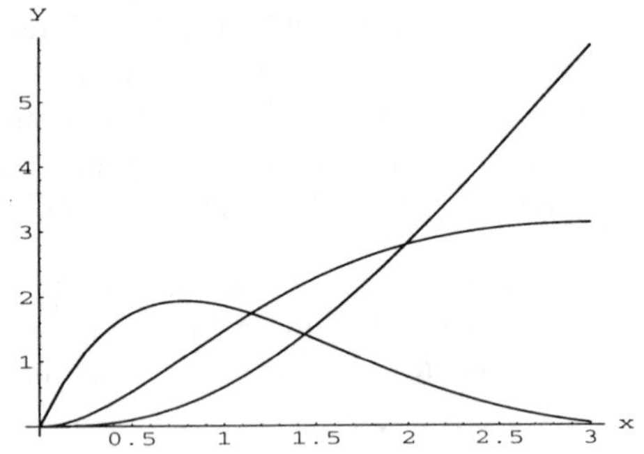

2. Approximate, to one decimal place, the instantaneous rate of change of the function $f(t) = 5^{\sqrt{\ln t}}$ at $t = 2$. Explain your answer in a manner that someone unfamiliar with calculus can understand. Write your answer in complete sentences that form a coherent paragraph or paragraphs.

Final Examination for Calculus 1: Test 6

1. For parts (a), (b), and (c), circle the correct answer and *state how you arrived at your answer.*

   (a) An antiderivative for the function $f(x) = \ln x$ is $F(x) = x \ln x - x + 1$.

   TRUE **FALSE**

   (b) If $f(t)$ is a quadratic function, then Simpson's method gives exact answers when used to approximate $\int_a^b f(t)\,dt$ no matter how many subdivisions are used.

   TRUE **FALSE**

   (c) If $f(t)$ is a quadratic function, then, if trapezoids are used to approximate $\int_a^b f(t)\,dt$ the answer will be exact, no matter how many subdivisions are used.

   TRUE **FALSE**

2. Find $f(x)$ if $f''(x) = 3e^x + 5\sin x$, $f(0) = 1$, and $f'(0) = 2$.

3. (a) Show that the function $y = \cos^2 x$ has a minimum and also two points of inflection in the interval $[0, \pi]$. Find those points and sketch the graph.

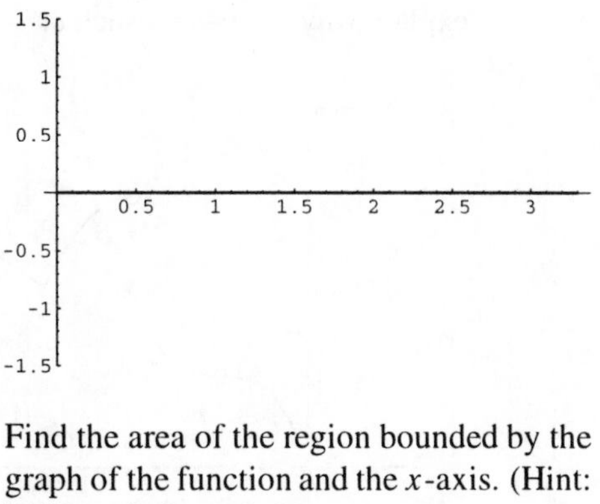

   (b) Find the area of the region bounded by the graph of the function and the x-axis. (Hint: Half-angle formula.)

4. Show that the area between the curve $y = 1/x$ and the x-axis from $x = 10$ to $x = 20$ is the same as the area between the curve and the x-axis from $x = 1$ to $x = 2$. Find another interval along the x-axis where the area between the curve and the interval is the same.

5. Suppose you are using Newton's method to find a root of $f(x) = 0$ for the function given in the graph below. If $x_1 = 2$ is your first approximation, indicate on the graph where (approximately) your second and third approximations x_2 and x_3 will be located. Draw lines and write a few sentences to explain how you find x_2 and x_3.

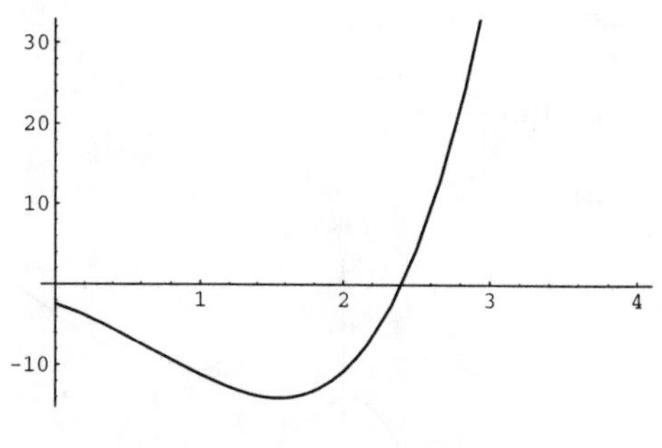

6. A playing field is to be built in the shape of a rectangle plus a semicircular area at each end. A 400 m race track is to form the perimeter of the field. Find the dimensions of the field if the rectangular part is to have as large an area as possible. Draw a picture, show all your work and **verify** that the dimensions that you find maximize the rectangular area. (In other words, there has to be an argument why you have a maximum rather than a minimum.)

7. A Flying Tiger is doing a nose dive along a parabolic path having the equation $y = x^2 + 1$, where x and y are measured in feet. The distance from the plane to the ground is decreasing at the constant rate of 100 ft/sec. How fast is the shadow of the plane moving along the ground when the plane is 2501 feet above ground? Assume that the sun's rays are vertical.

Final Examination for Calculus 1: Test 7

1. For each of the following, find the limit if it exists. If the limit does not exist, explain why it does not exist.

   (a) $\lim_{x \to \sqrt{2}} \frac{x^2 - 2}{x - \sqrt{2}}$

   (b) $\lim_{x \to 0} \frac{\sin(3x)}{\pi x}$

   (c) $\lim_{x \to \infty} \frac{3x^2 + 2x}{17x^2 + 5}$

2. For each of the following, find the limit if it exists. If the limit does not exist, explain why it does not exist.

   (a) $\lim_{x \to 1} \frac{x - 1}{|x - 1|}$

   (b)

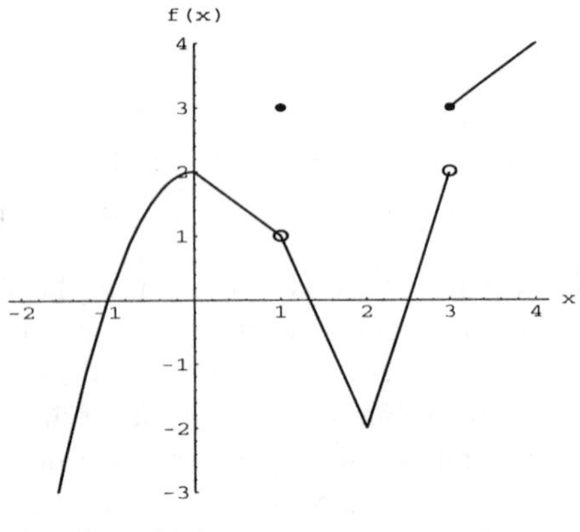

   (i) $\lim_{x \to -1} f(x) =$

   (ii) $\lim_{x \to 1} f(x) =$

   (iii) $\lim_{x \to 2^+} f(x) =$

   (iv) $\lim_{x \to 3} f(x) =$

3. For each of the following, find the indicated derivative.

   (a) $f(x) = 5x^{3/2} + \frac{7}{x} + 14$, find $f'(x)$.

   (b) $y = \frac{3x^2 + 2}{\sin(x)}$, find dy/dx.

   (c) $y = 2\cos(x/2)$, find d^2y/dx^2.

4. For each of the following, find the indicated derivative

   (a) $x^2 y = 3y + x\ln(y)$, find dy/dx when $x = \sqrt{3}$.

   (b) $h(x) = f\big(g(3x)\big)$, find $h'(2)$ if $g(6) = 1$, $f(1) = 4$, $f'(1) = 1/2$, and $g'(6) = 3$.

5. (a) If $\int_0^k (2kx - x^2)\,dx = 18$, find k.

   (b) Find all antiderivatives (or the indefinite integral) of $5\sin(3x)$.

   (c) Find all antiderivatives (or the indefinite integral) of $x/\sqrt{3x^2 + 5}$.

6. (a) Find all antiderivatives (or the indefinite integral) of $2/(3x + 1)$.

   (b) Find all antiderivatives (or the indefinite integral) of $(2x + 4)^3$.

   (c) If $\int_1^{10} f(x)\,dx = 4$ and $\int_3^{10} f(x)\,dx = -7$, then what is $\int_1^3 f(x)\,dx$?

7. (a) $f(x) = \begin{cases} 3x + 1 & x \le 2 \\ cx^2 & x > 2 \end{cases}$

   Is there a value of c for which f is continuous on $(-\infty, \infty)$? If so, find it. If not, explain why there is no such c.

   (b)

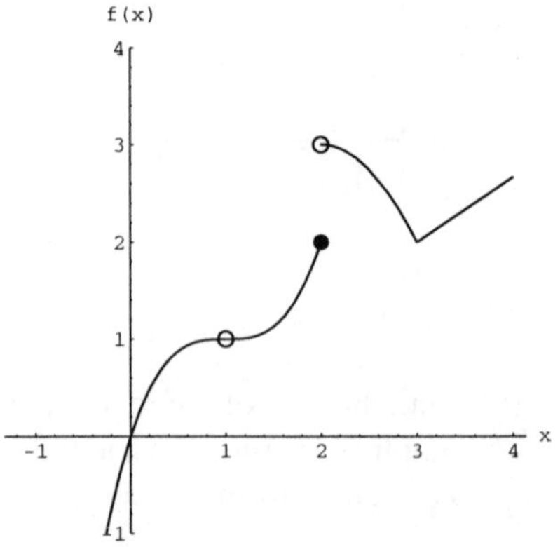

   (i) For which values of x is f discontinuous?

   (ii) For which values of x is f *not* differentiable?

8. (a) Find $\lim_{h \to 0} \dfrac{f(3+h) - f(3)}{h}$ if $f(x) = 2x^2 + 1$.

(b) Find $\lim_{h \to 0} \dfrac{3\sin\left(\frac{\pi}{6} + h\right) - \frac{3}{2}}{h}$.

9.

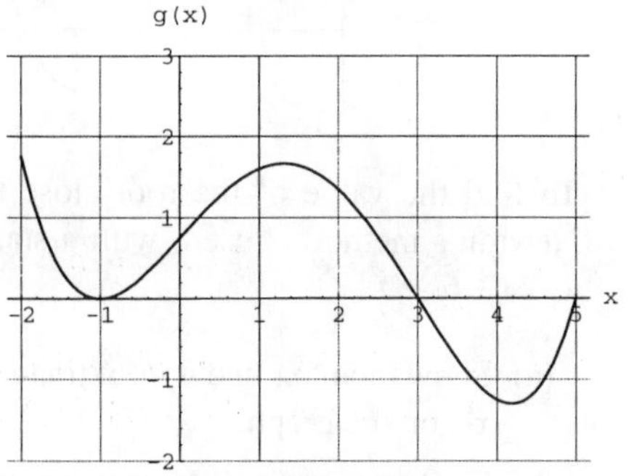

(a) Show that $\int_{-1}^{2} g(x)\,dx$ is between 2 and 5.

(b) For which value of c in $[-1, 5]$ will $\int_{-1}^{c} g(x)\,dx$ be the largest? Justify your answer.

10. Let $F(x) = \int_1^x \sqrt{t^2 + t}\,dt$.

(a) Find $F'(x)$.

(b) Find $\lim_{x \to 1} F(x)$.

11. A function f that is continuous for all real numbers x has $f(3) = -1$ and $f(7) = 1$. If $f(x) = 0$ for exactly one value of x, then which of the following could be x? Justify your answer.

(a) -1
(b) 0
(c) 1
(d) 4
(e) 9

12. If $f'(x) = x^2 + x - 12$, then f is increasing on

(a) $(-4, 3)$
(b) $(-3, 4)$
(c) $(-\infty, -\frac{1}{2})$
(d) $(-\infty, -4)$ and $(3, \infty)$
(e) None of the above.

Justify your answer.

13. Let f be the function that is defined for all real numbers x and that has the following properties:

(i) $f''(x) = 24x - 18$
(ii) $f'(1) = -6$
(iii) $f(2) = 0$

(a) Write an expression for $f(x)$.

(b) Find the average value of f on the interval $1 \le x \le 3$.

14.

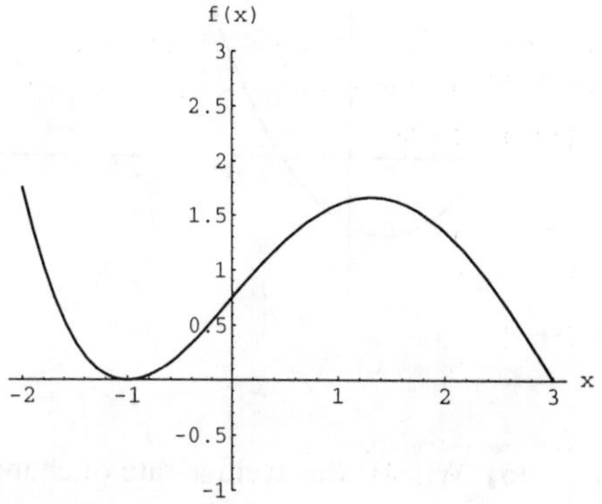

On which of the following intervals are both $dy/dx > 0$ and $d^2y/dx^2 < 0$?

I. $-2 < x < 0$
II. $0 < x < 1.5$
III. $1.5 < x < 3$

(a) I only
(b) II only
(c) III only
(d) I and II
(e) II and III

15. Suppose that $\lim_{x \to 3} f(x) = 7$. Which of the following must be true?

I. f is continuous at $x = 3$
II. $f'(3) = 7$
III. $f(3) = 7$

(a) None
(b) II only
(c) III only
(d) I and II only
(e) I, II, and III

16. Suppose that $f(1) = 0$ and that $1 \le f'(x) \le 2$ for x in $[0, 4]$. Use the Mean Value Theorem to explain why $f(4)$ cannot be 10.

17.

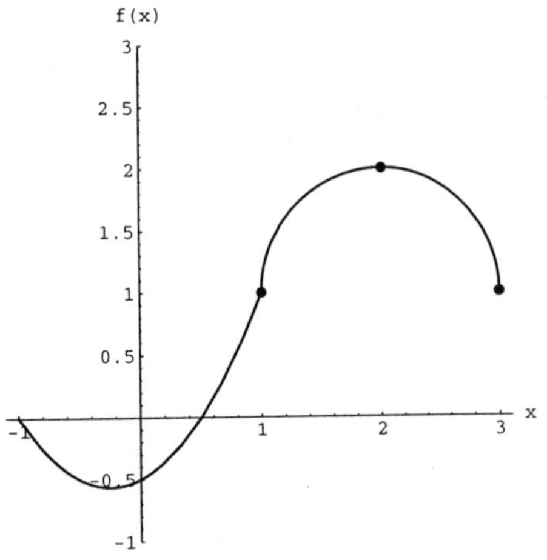

(a) What is the average rate of change of f on $[-1, 3]$?

(b) On what intervals is $f'(x)$ increasing?

(c) On what intervals is $f'(x)$ decreasing?

18. The volume of a cylindrical tin can with a top and bottom is to be 16π cubic inches. If a minimum amount of tin is to be used to construct the can, what must be the height, in inches, of the can? (You may want to know that the surface area of a cylinder, excluding a top and bottom, is $2\pi rh$.)

19.

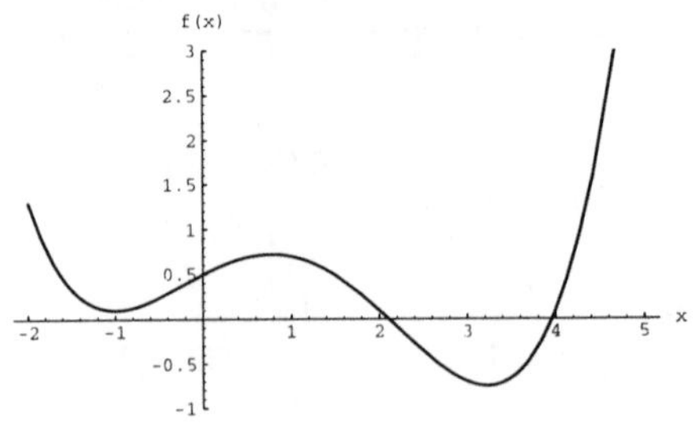

To find the value of the root close to $x = 2$, Newton's method is used, with a starting value $x_0 = 1.6$.

(a) Show the next two approximations, x_1 and x_2, on the graph.

(b) Explain what will happen as we continue the process.

20.

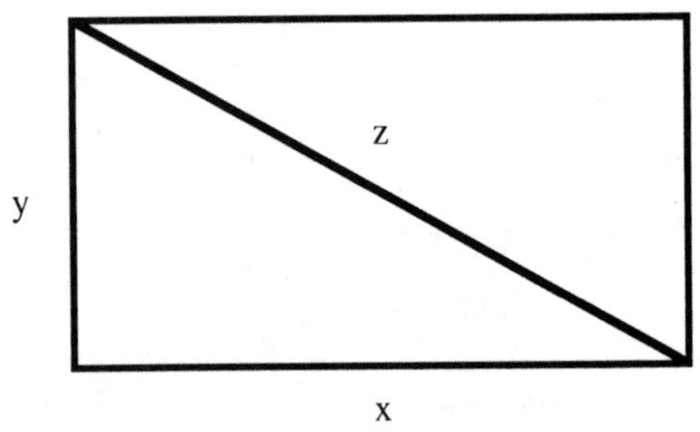

The sides of the rectangle above increase in such a way that $dz/dt = 1$ and $dx/dt = 3dy/dt$. At the instant when $x = 4$ and $y = 3$, what is the value of dx/dt?

Final Examination for Calculus 1: Test 8

Part I

This is the take-home (computer part of the final exam) worth 38 of the 150 points on the whole final. It is due at the beginning of the in-class portion. **You are to work alone on this exam.** *You may use Maple or any other computer or calculator software, and you may use the book, your notes, or any other course materials.*

1. Consider the initial value problem $y' = ay(1 - y/b)$, $y(0) = 10.0$. Use your student ID number to set the values of a and b. For a take the first three digits and divide by 1000; for b simply take the middle three digits. Exceptions: if either sequence of three digits begins with a zero, use a one instead of the zero. Example: for ID 015–46–0693, $a = 0.115$ and $b = 460$; for ID 247–08–1914, $a = 0.247$ and $b = 181$.

   (a) Produce an accurate graphical solution for this problem.

   (b) What is the long term behavior? (Note: you will need to do some experimenting to determine what "long term" means for your personal equation.)

   (c) At approximately what time is y increasing most rapidly?

   (d) Change the initial value to $y(0) = 5b/4$. How does the appearance of the solution change in this case? How do you explain the difference?

   (e) Using the same values of a and b as above, find x so that $f(x)$ is the global maximum of the function $f(x) = ax(1 - x/b)$ for $0 \le x \le \infty$. (You may use any combination of graphical or calculus methods, but explain very clearly what you are doing!) What connection, if any, does this problem have to (c) above? Explain!

2. A certain population $P(t)$ is measured at times $t = 1, 2, 3, 4, 5$. The values are 679, 1526, 3430, 7711, and 17,334. Your job is to figure out the population at times $t = 0$ and $t = 6$. Find a formula for $P(t)$ that agrees with this data and that allows you to calculate $P(0)$ and $P(6)$. (Hint: if you think P grows linearly you need to find b and m so that $P(t) = b + mt$; if you think P grows exponentially you need to find a and k so that $P(t) = ae^{kt}$; first decide which of these possibilities is more likely to be correct.)

3. Consider the initial value problem $y'(t) = \sin^2 t$, $y(0) = 0$.

   (a) Is $y(t) = \frac{1}{3}\sin^3 t$ a solution of this problem? Explain!

   (b) Produce a graphical solution for $y(t)$ on the domain $0 \le t \le 2\pi$, and give the value $y(2\pi)$ accurate to at least two decimal places.

   (c) For 5 bonus points, use the trig identity $\sin^2 t = \frac{1}{2}\big(1 - \cos(2t)\big)$ to get a formula solution for $y(t)$. No credit unless you also verify that your formula is correct!

Part II

You have 30 minutes for this part of the exam, but you may go on to Part III as soon as you are finished with Part II. No calculators are allowed until Part II has been turned in; no books or notes are allowed during the entire exam.

1. Find formulas for the derivatives of the following functions. Answers should read *derivative* = formula where *derivative* is written with the appropriate f' or Leibniz notation.

   (a) $F = \pi r^2 + \dfrac{1}{5r} + 3^r$

   (b) $x = t^3 \cos(2t)$

   (c) $y = \dfrac{(\ln x)^4}{\tan x}$

   (d) $w = \ln \sqrt[3]{\dfrac{z(z-2)}{z+3}}$

2. Sketch the graphs and indicate the intercepts.

   (a) $y = \ln \dfrac{1}{x}$ (b) $y = (e^{-x})^4$

Part III

On this part of the exam no notes or books are allowed, but calculators are OK. Be sure to show your work for full credit.

1. When a wild new fashion in clothes (or hair styles) is introduced, it spreads slowly through the population at first but then speeds up as more people become aware of it. Eventually, however, the pool of those willing to try new fashions begins to dry up, and while the number of people adopting the new fashion continues to increase, it does so at a decreasing rate. Later, the fashion goes "out" and disappears very quickly, although a few stragglers never give it up at all. Sketch a graph of the number of people who wear the fashion as a function of time.

2. Find a formula for $F(t)$ if $F'(t) = 21t^{4/3} - t^2 + 10/3$ and $F(1) = 7$.

3. Suppose f is a differentiable increasing function with domain $0 \le x \le \pi$, and range $-1 \le x \le 1$. Let g be the inverse function. The domain of g is ____. If $f(\pi/3) = -1/2$, then $g($____$) =$ ____. If $f'(\pi/3) = \sqrt{3}/2$, then $g'($____$) =$ ____.

4. Suppose f is locally linear at $x = -2$, $f(-2) = 1$ and $f'(-2) = -3$. Estimate $f(-2.4)$.

5. The volume of a pyramid of height h on a square base with sides of length s is $V = \frac{1}{3}s^2h$.

   (a) Compute the partial derivatives $\partial V/\partial h = V_h$ and $\partial V/\partial s = V_s$, and write the total microscope equation for V.

   (b) Suppose the length of the base s is more than the height h. If s decreases by 0.25 cm and h increases by 0.5 cm, does V increase or decrease? Why?

   (c) Express the *relative error* in V in terms of the relative errors in s and h.

6. For each function whose graph $y = f(x)$ is sketched below, sketch the graph of the derivative function $y = f'(x)$ on the same set of axes.

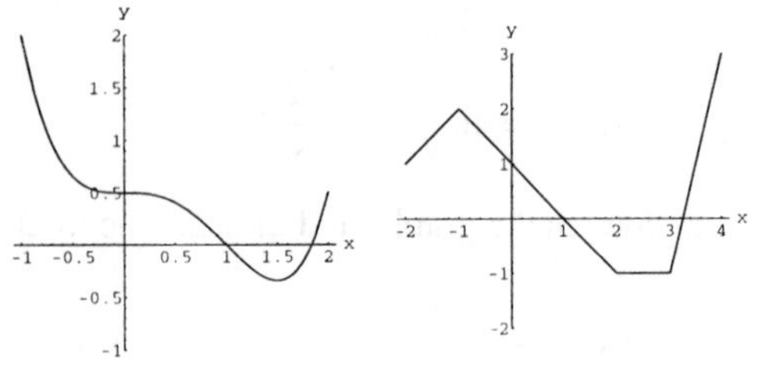

7. The graph of $f'(x)$ is shown below. If $f(0) = 2$, and $f(x) \to -1$ as $x \to -\infty$, sketch the graph of $f(x)$ on the same set of axes.

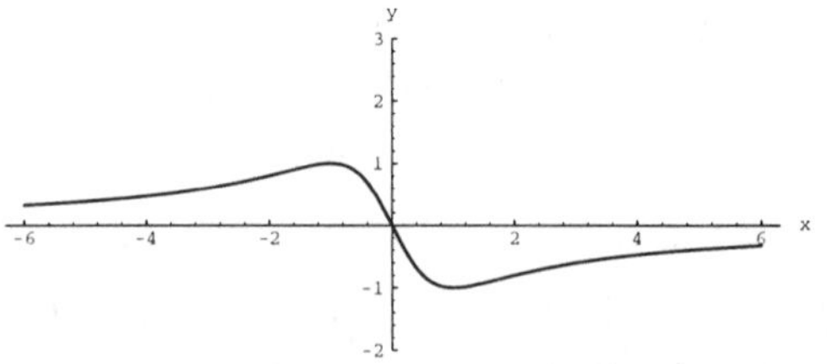

8. Write an equation for the tangent line to the graph $y = 6\sqrt{x+1}$ at $x = 8$.

9. Use the graph of $y = h(x)$ given here to produce good sketches of the modified graphs. Be sure to put enough numbers on your axes so that all the graph's features are clearly labeled.

   (a) Sketch $y = h\left(\frac{1}{2}x\right)$.

   (b) Sketch $y = 3h(x)$.

   (c) Sketch $y = h(x-1)$.

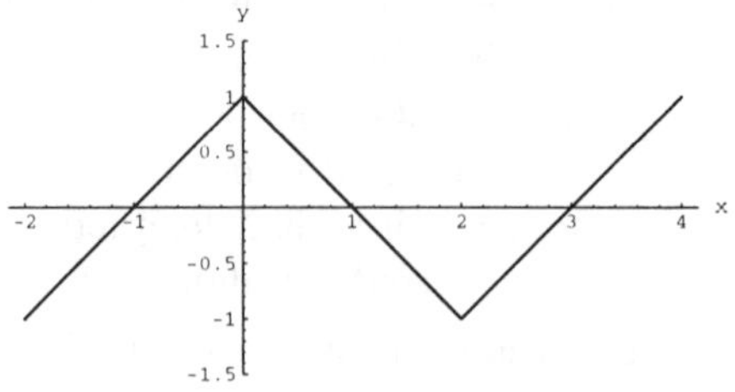

10. The population of a highly desired, but rare, species of salmon is increasing at an annual rate equal to 5% of the current population P (which is measured in millions); this takes into account the birth rate and the death rate due to natural causes such as disease and predation by animals other than humans. Meanwhile, fishermen catch 2.8 million of these salmon each year.

    (a) Write a differential equation that describes the rate of change of this salmon population.

    (b) If the current population is 100 million, use Euler's method with three steps to approximate the population three years from now.

11. Consider the function $y = f(x) = xe^{-x} = x/e^x$.

    (a) What is the natural domain of this function?

    (b) Describe the behavior of $f(x)$ as $x \to \infty$. Use words and/or a graph.

    (c) Find all critical points of f.

    (d) Where is f decreasing? Where is f increasing?

    (e) What kind of extreme point, if any, does this function have at each critical point: local maximum? local minimum? also global max or min? Briefly explain.

12. Consider the function

$$y = g(x) = \frac{2x^3}{x^2+1} = 2x - \frac{2x}{x^2+1}.$$

We observe that

$$g'(x) = \frac{2x^2(x^2+3)}{(x^2+1)^2} \quad \text{and} \quad g''x) = \frac{4x(3-x^2)}{(x^2+1)^3}.$$

    (a) Describe the behavior of $g(x)$ (i.e., precisely *how* $g(x) \to \infty$ or $g(x) \to -\infty$ as $x \to \infty$ and also as $x \to -\infty$. Use words and/or a graph.

    (b) Find all critical points of g.

    (c) What kind of extreme point, if any, does this function have at each critical point: local maximum? local minimum? also global max or min? Briefly explain.

13. A wooden crate is to have a capacity of 343/2 cubic inches. It has vertical sides, a square bottom, and is open on top.

    (a) Write a formula for the total surface area A of the crate in terms of one independent variable.

    (b) Give a domain for A that makes sense for this problem.

    (c) Find the dimensions that require the least amount of wood in the construction (i.e., so that the surface area is minimized). Explain your answer adequately.

Final Examination for Calculus 1: Test 9

This test was written to be taken in two hours. You may have a full three hours to complete it.

Hand Calculations

1. Derivatives

   (a) If $f(x) = \sin(x)$, then $f'(x) =$

   (b) If $f(x) = \sin(x^2)$, then $f'(x) =$

   (c) If $f(x) = \cos(x)$, then $f'(x) =$

   (d) If $f(x) = \cos(e^x)$, then $f'(x) =$

   (e) If $f(x) = xe^x$, then $f'(x) =$

   (f) If $f(x) = e^{-2x^2}$, then $f'(x) =$

   (g) If $f(x) = e^{\cos(x)}$, then $f'(x) =$

   (h) If $f(x) = \ln(x)$, then $f'(x) =$

   (i) If $f(x) = x\ln(x)$, then $f'(x) =$

   (j) If $f(x) = x^2$, then $f'(x) =$

   (k) If $f(x) = 4x$, then $f'(x) =$

   (l) If $f(x) = 6x^5$, then $f'(x) =$

2. Evaluate each of the following integrals.

   (a) $\int_0^2 x^2\,dx$

   (b) $\int_0^2 x^4\,dx$

   (c) $\int_a^t e^x\,dx$

   (d) $\int_0^1 e^{-x}\,dx$

   (e) $\int_0^1 e^{-2x}\,dx$

   (f) $\int_0^{\pi/2} \cos(x)\,dx$

   (g) $\int_0^{x/2} \cos(2x)\,dx$

   (h) $\int_0^{\pi} x\sin(x^2)x\,dx$

   (i) $\int_0^{\pi/2} e^{\sin(x)}\cos(x)\,dx$

   (j) $\int_1^2 (1/x)\,dx$

   (k) $\int_0^3 x\cos(x^2)x\,dx$

   (l) $\int_0^3 dx = \int_0^3 1\,dx$

3. Misconceptions:

   (a) Writing $\int_0^3 t\,dt = \dfrac{3^2}{2} - \dfrac{0^2}{2} + C$ would be a sure sign of calculus illiteracy. Why?

   (b) Writing

   $$\int_0^t \text{Sin}(3x)\,dx = \text{Sin}\left(\frac{3t^2}{2}\right) - \text{Sin}(0)$$

   would be a sure sign of calculus illiteracy. Why?

4. Write down the formula for the solution of the differential equation: $y'(x) = 0.4y(x)$ with $y(0) = 8$.

Calculus Ideas

1. For a function $f(x)$, the function $f'(x)$ is another function that measures what quantity?

2. What happens to the plots of the $\dfrac{e^{x+h} - e^x}{h}$ curves as h closes in on 0? Explain!

3. Sketch below the graph of a function $f(x)$ such that $f'(x)$ is positive for all x.

4. If $f'(x)$ is positive for all x with $0 \le x \le 1$, then which is the larger: $f(0)$ or $f(1)$?

5. Sketch the graph of a function $f(x)$ such that $f'(x)$ is positive for all x with $0 \le x < 1/2$, and $f'(x)$ is negative for $1/2 < x < 2$. Where is the crest?

6. Sketch on the axes below the graph of a function $f(x)$ such that $f(0.5) = 1$ and $f'(x) = 0$ for all x.

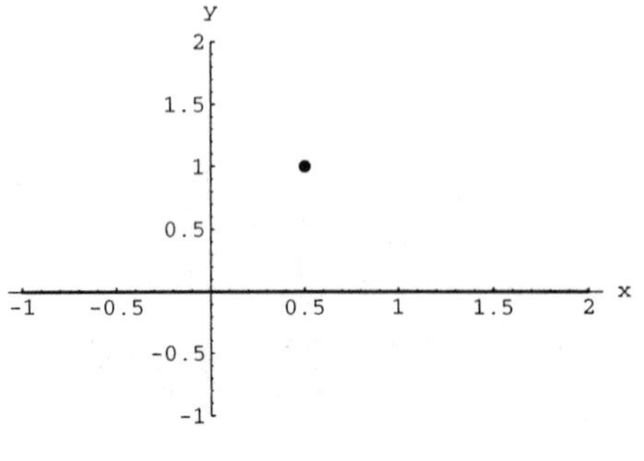

7. Here are plots of x and $\sin x$ for $0 \le x \le 2$:

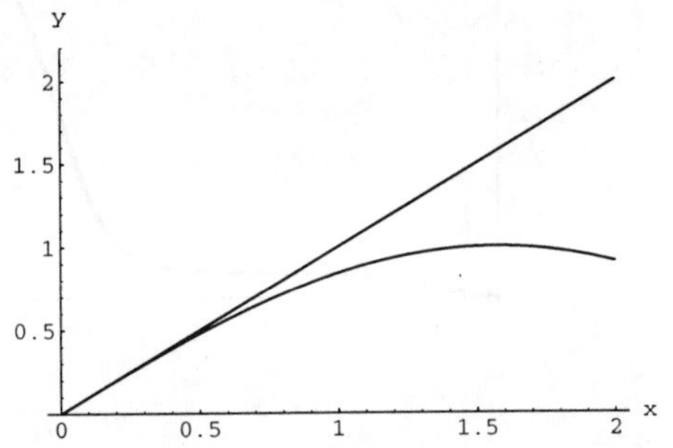

Use the Race Track Principle to explain why $x \ge \sin x$, for all $x \ge 0$. Remember that $-1 \le \cos x \le 1$ no matter what x is.

8. Here are plots of the solutions of the differential equations $y'(x) = 0.6y(x)$ with $y(0) = 2$ and $y'(x) = 1.2y(x)\left(1 - \dfrac{y(x)}{100}\right)$ with $y(0) = 2$.

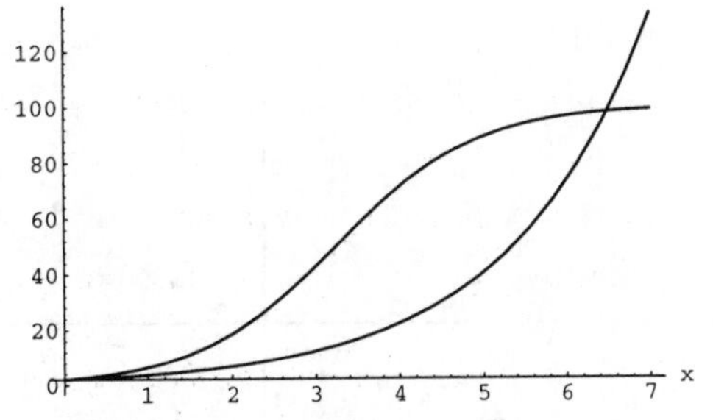

(a) Notice that the plots start out together but share very little ink. Why do you think this happened?

(b) Now look at plots of the solutions of the differential equations $y'(x) = 0.6y(x)$ with $y(0) = 2$ and $y'(x) = 0.6y(x)\left(1 - \dfrac{y(x)}{100}\right)$ with $y(0) = 2$.

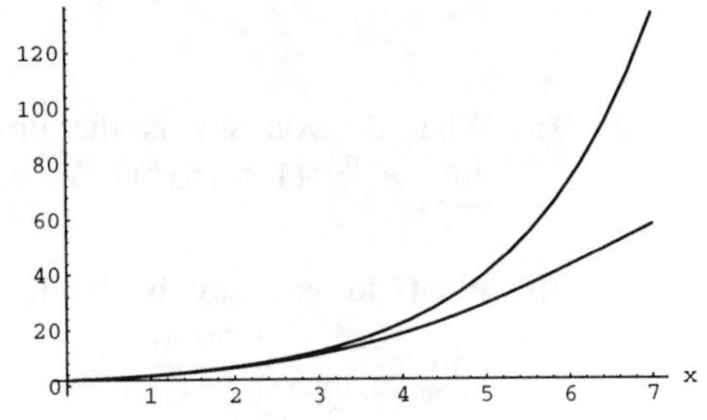

Explain why, as x advances from 0, the plots of both of these solutions had no choice but to share a lot of ink initially. Explain why for larger x, the plots had no choice but to pull apart, with one plot eventually sailing way above the other.

9. Here's a plot of $f(x) = 4(x^3 + 1)e^{-x^2}$.

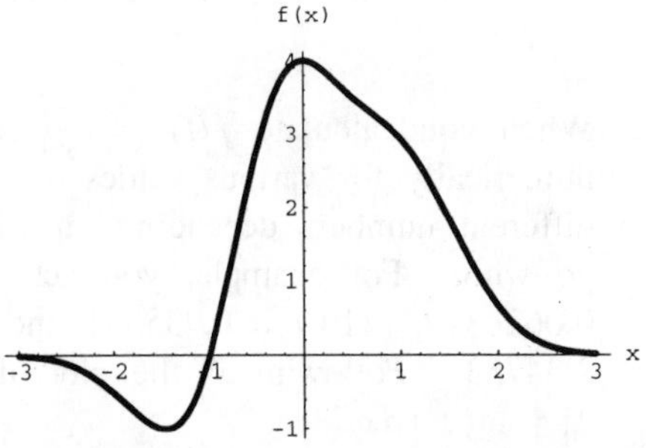

Look at the plot and then use the words positive, negative, or zero to complete the blanks:

(a) $\displaystyle\int_{-3}^{3} f(x)\,dx$ is ________________.

(b) $\displaystyle\int_{0}^{3} f(x)\,dx$ is ________________.

(c) $\displaystyle\int_{-3}^{-1} f(x)\,dx$ is ________________.

(d) $\displaystyle\int_{-1}^{2} f(x)\,dx$ is ________________.

10. Here is the plot of a certain function $g(x)$.

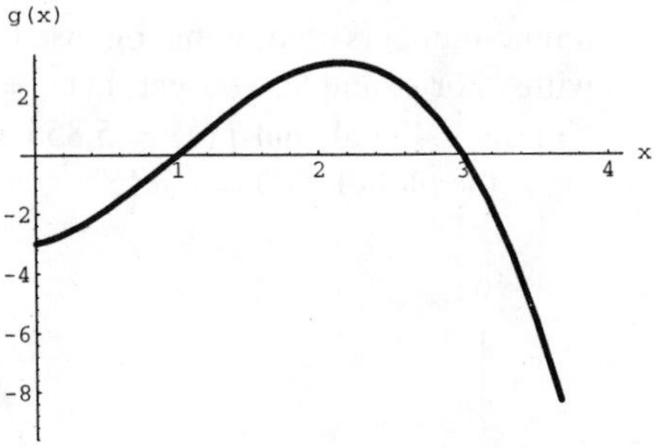

Explain how the plot of $g(x)$ signals that the plot of $f(t) = \int_0^t g(x)\,dx$

(a) goes down as t advances from 0 to 1;

(b) goes up as t advances from 1 to 3;

(c) goes down as t advances from 3 to 4.

11. Explain the idea behind the formula

$$\int_a^b f(x)\,dx = \int_a^c f(x)\,dx + \int_c^b f(x)\,dx$$

for any number c with $a < c < b$.

12. What's the idea behind integration by trapezoids?

13. When you calculate $f(t) = \int_0^t 4\sin(x^3)\,dx$ numerically for various values of t, you get different numbers depending on what t you go with. For example, you get $f(0.5) = 0.0624349$, $f(1.0) = 0.935381$, and $f(1.5) = 2.34753$. Following is the plot of $f(t) = \int_0^t 4\sin(x^3)\,dx$.

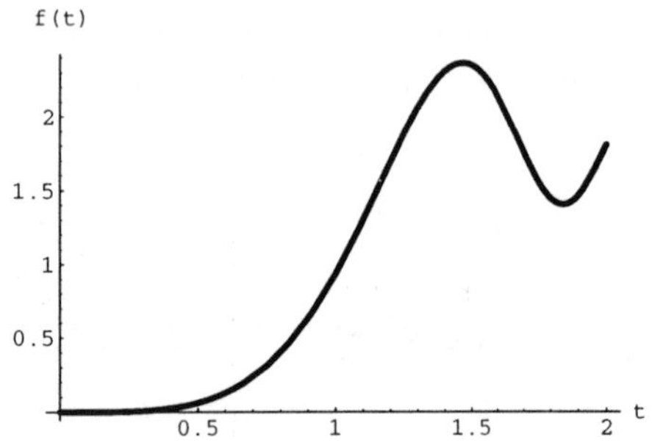

Even though you don't have a clean formula for $f(t)$, you can give a clean formula for $f'(t)$. Do it.

14. When you calculate $f(t) = \int_0^t e^{x\cos(2x)}\,dx$ numerically for various values of t, you get different numbers depending on what t you go with. For example, you get $f(1) = 1.12294$, $f(2) = 1.42524$, and $f(3) = 5.85576$. Following is the plot of $f(t) = \int_0^t e^{x\cos(2x)}\,dx$.

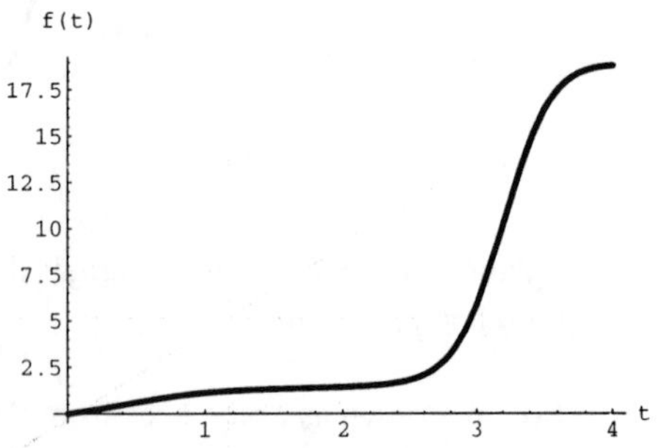

Now, here is a plot of the solution of the differential equation $y'(t) = e^{x\cos(2x)}$ with $y(0) = 0$.

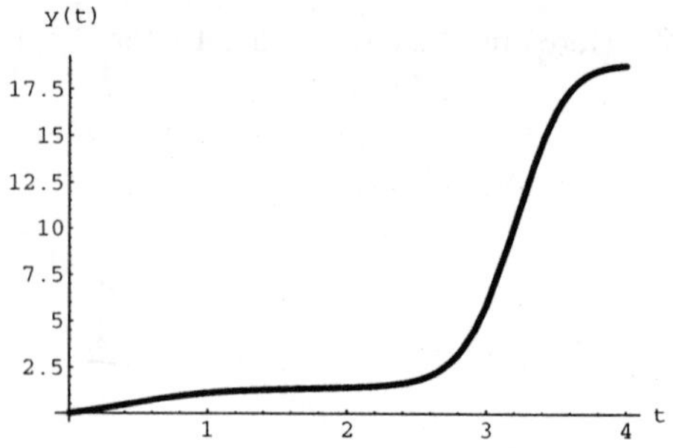

Explain why it is no accident that the plots are dead ringers for each other.

Calculus Literacy

1. Here are some data points:

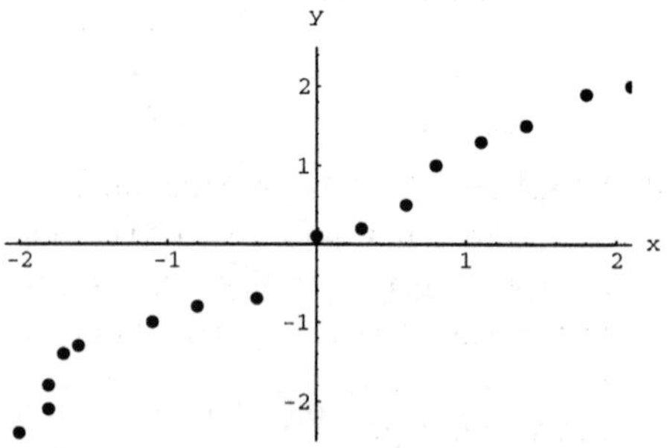

Sketch in the plot of a function, $f(x)$, like Mathematica's `Interpolation` function, that goes through all the plotted points and goes with the flow. Do you think that it is plausible to say that $f'(x) \geq 0$ for $-2 \leq x \leq 2$? Explain.

2. (a) What do you say is the limiting value $\lim_{x\to\infty} e^{-0.1x}(1 + 15x^8)$?

   (b) What do you say is the limiting value $\lim_{x\to\infty} \dfrac{6e^{0.03x} + 50\sin x}{3e^{0.03x} + 25\sin x}$?

   (c) Why do lots of folks say that exponential growth is awesome?

3. (a) Use the axes below to give hand sketches of the plots of $f(x) = e^x$ and $g(x) = e^{-x}$. Each of the plotted points is on at least one of the plots. Label each curve.

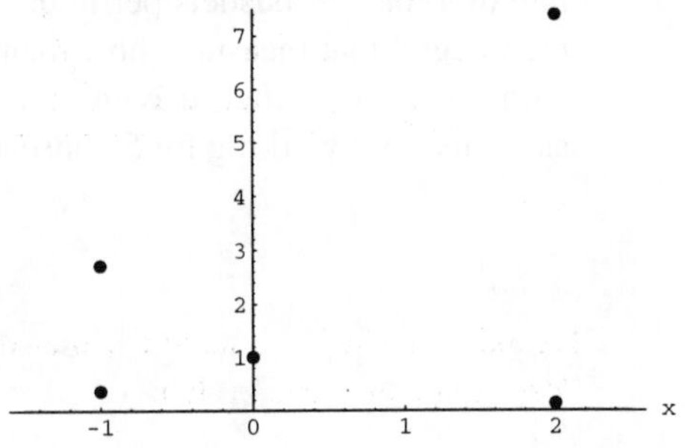

(b) Use the axes below to give hand sketches of the plots of $f(x) = \sin x$ and $g(x) = \cos x$. Each of the plotted points is on at least one of the plots. Label each curve.

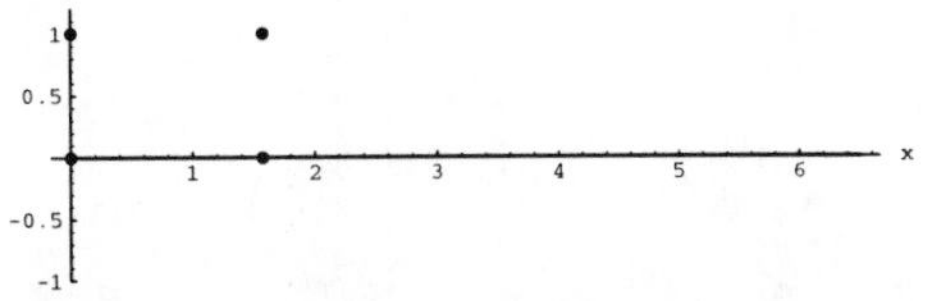

4. Someone says, "The line functions $f(x) = ax + b$ are those that grow by a fixed amount every time x goes up by a fixed increment." Another person says, "The exponential functions $f(x) = ae^{m}$ are those that grow by a fixed percentage every time x goes up by a fixed increment." Another person exclaims, "Exponential growth is awesome!" Are any of these folks right? If so, then which of them is (are) right?

5. From the Money section of the newspaper *USA Today*, June 18, 1992: "Rates on credit card applications don't take into account monthly, and sometimes daily, compounding of finance charges. The average rate on credit cards is 18.5%, but most consumers pay effective rates of 20% or more because of compounding methods." What does this mean?

6. Good representative plots of functions exhibit all the dips and crests of the graph and give a strong flavor of the global scale behavior. If your plot of a function includes all points at which the derivative is 0, explain why you can (or cannot) be sure that your plot does not miss any of the dips and crests of the graph of the function.

7. Here are plots of four functions. Two of them are plots of the derivatives of the functions plotted in the other two. Match the plots of the functions with the plots of their derivatives.

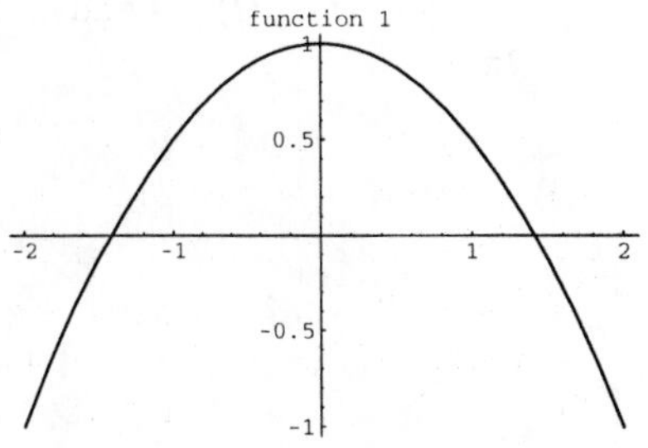

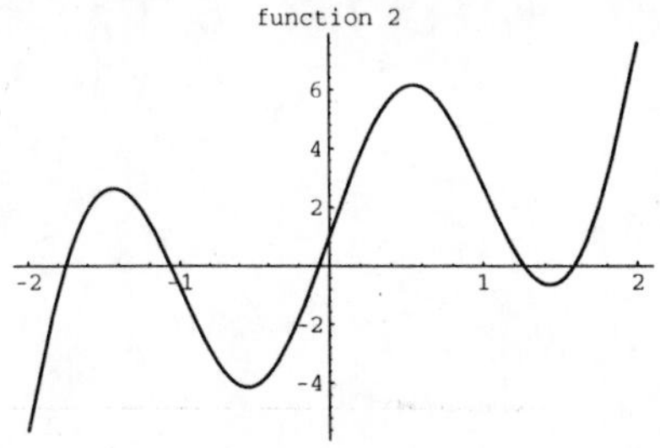

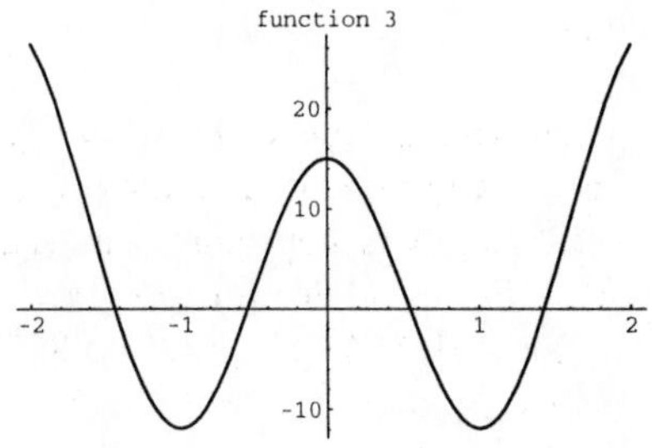

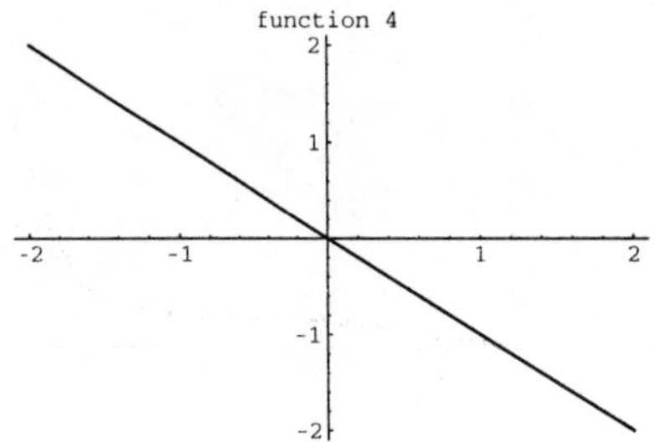

8. Let $x(t) = \cos t$ and $y(t) = \sin t$.

   (a) Sketch the curve traced out by $\{x(t), y(t)\}$ as t advances from 0 to $\pi/2$.

   (b) Sketch the curve traced out by $\{x(t), y(t)\}$ as t advances from 0 to π.

   (c) Sketch on the axes below the curve traced out by $\{x(t), y(t)\}$ as t advances from 0 to 2π.

9. When you are fresh, you find that you can harvest 150 bushels of corn per hour. But as the day wears on, your efficiency is somewhat decreased. In fact, after t hours from the beginning of the day, you find that you are harvesting at a rate of $150e^{-0.5t}$ bushels per hour. Write down the integral that measures how many bushels of corn you harvest after arriving in the field fresh as a daisy and working for 5 consecutive hours.

Final Examination for Calculus 1: Test 10

1. Find a solution $y(t)$ to each initial value problem, then verify that your solution is correct. (Be sure to check both the initial value and the differential equation.)

   (a) $\dfrac{dy}{dt} = 6t^2 - 5e^{5t} + \dfrac{6}{t^3} - 7, \quad y(1) = 4$

   (b) $\dfrac{dy}{dt} = -0.03y, \quad y(0) = 12$

2. There is no formula for the antiderivative of the function $\sin(t^3 - 4)$, so in order to estimate a solution $y(t)$ to the initial value problem

$$y' = \sin(t^3 - 4), \quad y(0) = 5,$$

   we must use Euler's method.

   (a) Modify the program below so that it estimates the graph of $y(t)$ on the interval $0 \le t \le 10$ using a time step of $\Delta t = 0.01$. You may assume that $4 \le y \le 6$ over the given interval.

```
! SIRPLOT

SET WINDOW 0,75,50,1500
LET tinitial = 0
LET tfinal = 75
LET t = tinitial
LET y = 100
LET numberofsteps = 150
LET deltat=(tfinal-tinitial)/numberofsteps
FOR  k = 1  TO  numberofsteps
     LET yprime = 0.1*y*(1 - y/1000)
     LET deltay = yprime*deltat
     PLOT t, y ; t+deltat, y+deltay
     LET t = t + deltat
     LET y = y + deltay
NEXT k
END
```

   (b) How could you further modify the program to get an even better estimate of the graph of $y(t)$?

   (c) Find a value of t, $0 \le t \le 10$, for which the *slope* of $y(t)$ is 0. Give your answer to 4 digits after the decimal point. Be sure to explain your method.

3. The following differential equation describes the rate of change of a chipmunk population C.

$$C' = -0.1C\left(1 - \frac{C}{10000}\right)\left(1 - \frac{C}{500}\right) \text{ chipmunks per year}$$

   (a) Find all values of C for which the chipmunk population neither increases nor decreases.

   (b) At the time when the chipmunk population is $C = 4500$, is the chipmunk population increasing or decreasing? Explain.

   (c) Suppose the chipmunk population is $C = 4500$ today ($t = 0$). Estimate the chipmunk population one year from now using a time step of $\Delta t = \frac{1}{2}$ year.

4. The susceptible, infected, and recovered populations (S, I, and R, respectively) for a rather short-lived epidemic of a disease similar to measles are graphed below. They were graphed using the differential equations

$$\begin{aligned} S' &= -0.00002SI \\ I' &= 0.00002SI - \tfrac{1}{20}I \\ R' &= \tfrac{1}{20}I \end{aligned}$$

   and the initial values $S = 56600$, $I = 5500$, and $R = 2300$.

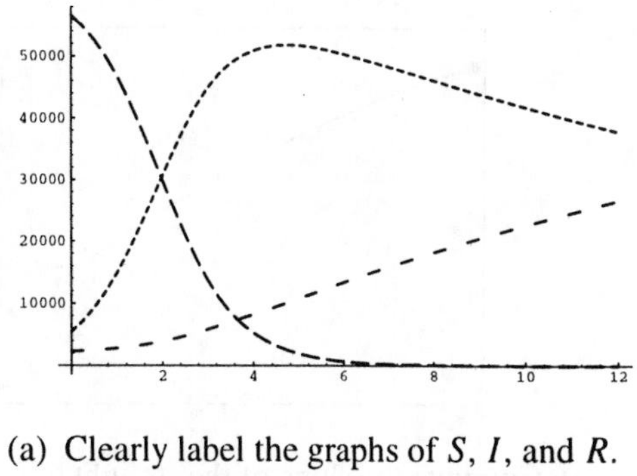

   (a) Clearly label the graphs of S, I, and R.

   (b) After approximately how many days does the infection hit its peak? Approximately how many people are infected at that time? Mark the peak of the infection on the appropriate graph.

(c) Mark the threshold value for S on the appropriate graph, then find the (exact) threshold value for S for this epidemic. Finally, explain how *and why* knowing the threshold value for S could be used to prevent a recurrence of this epidemic in the future.

5. A fungus is growing *at a rate proportional to its current size*. Suppose the fungus weighs 12 grams at time $t = 0$ hours.

   (a) Write an initial value problem (differential equation *and* initial value) for the weight F of the fungus.

   (b) Write the solution to your initial value problem from part (a). That is, write a formula for the weight F of the fungus at time t hours.

   (c) Use that the fungus weighs 15 grams after $t = 6$ hours to complete your formula from part (b).

   (d) How much does the fungus weigh after $t = 10$ hours?

   (e) How many hours does it take for the fungus to reach 24 grams in weight?

6. Consider the function $f(x) = \sin(x)$.

   (a) Compute $f'(2)$. Be sure to set your calculator in radian mode. Report your answer to 6 digits after the decimal point.

   (b) In the following picture of a MICROSCOPE window, fill in the coordinates of the endpoints.

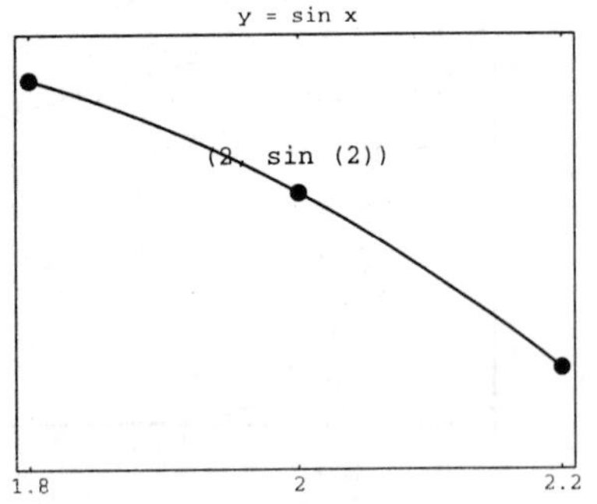

   (c) Compute the slope of the straight line connecting the endpoints in the picture above.

   (d) Compute the slope of the straight line connecting the point (2, sin(2)) and the right endpoint in the picture above.

   (e) Which of the two slopes that you computed in parts (c) and (d) is a better estimate of $f'(2)$? Use the picture in part (b) to explain why this should be so.

7. Compute the derivative $f'(x)$ of each function $f(x)$. Simplify your answers enough so that there are no negative or fractional exponents.

   (a) $f(x) = 4x^{11} - \frac{3}{2}x^2 + \frac{5}{x^2} + \ln(4x^2 + 3)$

   (b) $f(x) = 3^x - 3e^{2x} + \sqrt{x^4 + 3x^2 + 4}$

   (c) $f(x) = \left(4x^3 + \sin(2x)\right)^5$

   (d) $f(x) = 7\sqrt{x}\cos(3x + 4)$

8. The graph of the function $f(x) = x^4 + 2x^3 - 1$ is shown below.

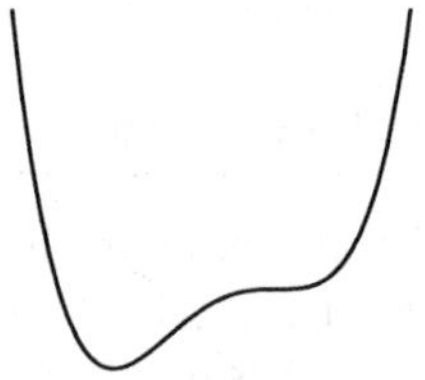

   (a) Find the critical points for $f(x)$. Be sure to find both the x-coordinate and the y-coordinate of each point and to show your work. Label these points on the graph with their x and y coordinates. [HINT: There are two critical points.]

   (b) Determine which critical points (if any) are local maximum points and which are local minimum points for $f(x)$. Which (if any) are global maximum points and which are global minimum points? Label this (these) point(s) on the graph.

   (c) Find the inflection points for $f(x)$. Be sure to find both the x-coordinate and the y-coordinate of each point and to show your work. Label these points on the graph with both their x and y coordinates.

   (d) Sketch in the x-axis and the y-axis on the graph. [HINT: What is the y-intercept for the graph?]

9.

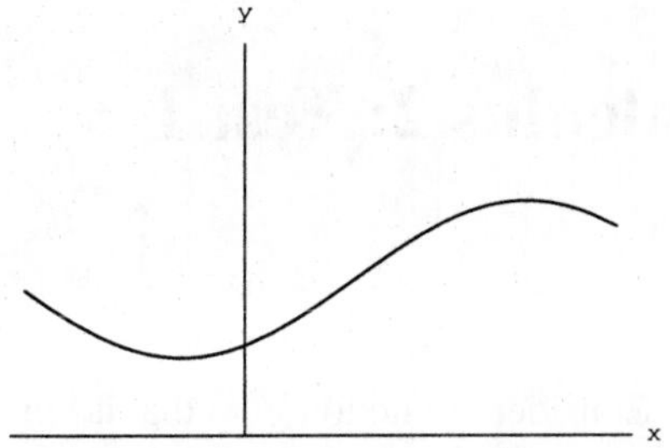

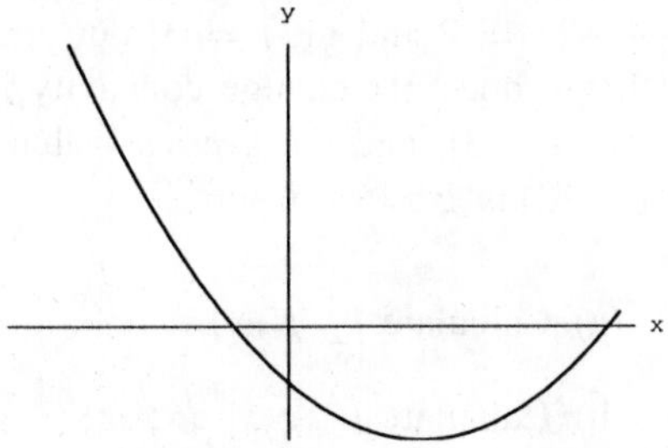

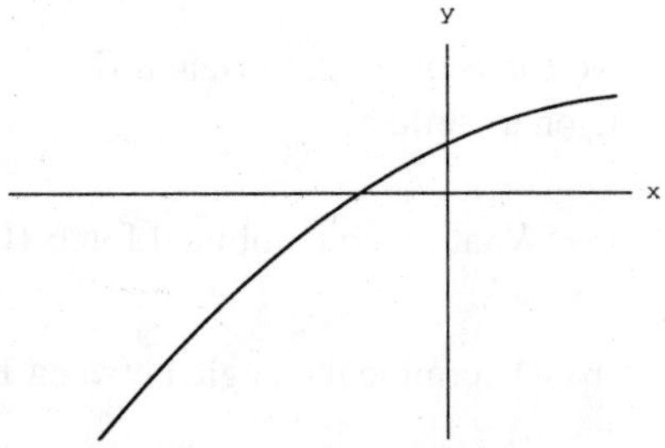

(a) For which one of the graphs above is it true that $f'(x) > 0$ for every x? (That is, for which function f is $f'(x)$ always positive?)

(b) For which one of the graphs above is it true that $f''(x) > 0$ for every x? (That is, for which function f is $f''(x)$ always positive?)

10. You have been asked to design a 1-liter (1000 cm^3) oil can shaped like a right circular cylinder (that is, shaped like a can!) with the smallest possible surface area. (The can with the smallest possible surface area will require the least material to build, hence should cost the least to manufacture.)

(a) Let r be the radius of the can (in cm) and let h be the height of the can (in cm). Then the surface area A (in cm^2) and the volume V of the can (in cm^3) are given by

$$A = 2\pi r^2 + 2\pi rh; \quad V = \pi r^2 h = 1000.$$

Use the equation $\pi r^2 h = 1000$ to rewrite the area formula given above as

$$A = 2\pi r^2 + \frac{2000}{r}.$$

(b) The graph of the surface area function $A = 2\pi r^2 + 2000/r$ is shown below. Find the radius r that yields the smallest possible (minimum) surface area. (You should give both an exact answer and an answer accurate to two digits after the decimal point.) Then find the smallest possible surface area. Label both coordinates of the corresponding point on the graph of A.

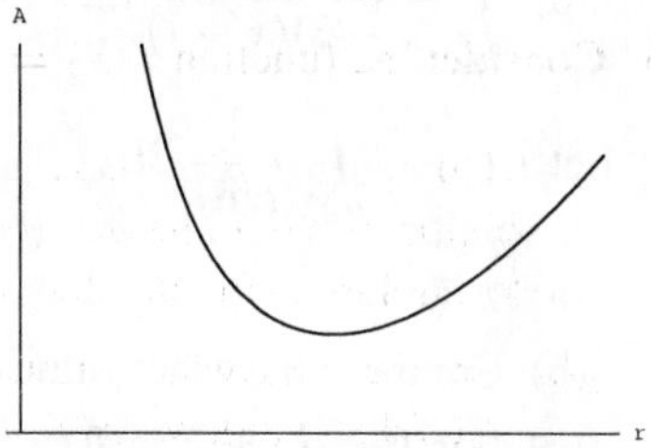

(c) Use the equation $\pi r^2 h = 1000$ to find the height of the can with minimum surface area. What is the relationship between the values of r and h that yield minimum surface area? What is the relationship between the *diameter* and the *height* of the can with minimum surface area? Provide a sketch of this can.

Final Examination for Calculus 2: Test 1

On this exam no books are allowed, but calculators are OK for numerical work. For integration, exact summation of series, and so forth, you MUST SHOW your work for full credit. You may bring a reference card with at most eight formulas written on it. There are 150 points.

1. Find a formula for $z(t)$ if $z'(t) = \sin 8t - 9t^2$ and $z(0) = -9/8$.

2. Compute the integral. Show your work!

   (a) $\displaystyle\int t \ln t \, dt.$

   (b) $\displaystyle\int_1^{\infty} \frac{1}{\sqrt{x}} \, dx.$

   (c) $\displaystyle\int \frac{1}{\sqrt{1-y^6}} y^2 \, dy.$

   (d) $\displaystyle\int \frac{2x}{(x-3)(x-1)} \, dx.$

3. Let $F(x) = \displaystyle\int_2^x \frac{1}{t \ln t} \, dt.$

   (a) $F'(x) =$.

   (b) Compute the exact value of $F(5)$.

   (c) For $1 < x < 2$, is $F(x)$ positive or negative?

4. Solve for $y(t)$ if $y' = t + ty^2$ and $y(0) = \sqrt{3}$.

5. A particle moves so that $\mathbf{r}(t) = 3\cos 2t\,\mathbf{i} + \sin 2t\,\mathbf{j} = (3\cos 2t, \sin 2t)$.

   (a) Compute the velocity vector $\mathbf{v}(t)$ and the speed.

   (b) Give an equation in the variables x and y that describes the path of the motion. Sketch the path for $0 \leq t \leq \pi$, and indicate the direction of motion.

6. A region D is enclosed by the y-axis, the positive x-axis, and the curve $y = \sqrt{x}e^{-x}$. By revolving this region around the x-axis we obtain a three-dimensional solid. Compute the exact volume of this solid.

7. Consider a function $g(x)$ that is continuous, and whose derivatives are continuous on the interval $[-1, 3]$. The only values that you know are $g(-1) = 2$ and $g(3) = 3$; but you also know that g does not change concavity in the interval $[-1, 3]$, and the average value of $g(x)$ on $[-1, 3]$ is $\bar{g} = 2.25$.

   (a) Calculate $\int_{-1}^{3} g'(x) \, dx$.

   (b) Calculate $\int_{-1}^{3} g(x) \, dx$.

   (c) Make a plausible sketch of $y = g(x)$.

8. Two forces $\mathbf{F} = (2, -5)$, and $\mathbf{G} = (5, 2)$ are acting on a particle.

   (a) What is the combined force $\mathbf{H}$ on the particle?

   (b) Determine the angle between $\mathbf{F}$ and $\mathbf{G}$.

   (c) The particle is displaced from the point $P(-1, 3)$ to the point $Q(3, 0)$. Compute the displacement vector $\mathbf{D}$.

   (d) Compute the work done by the force $\mathbf{F}$ as a particle is displaced along $\mathbf{D}$.

   (e) Compute the "effective force," i.e., the vector projection of $\mathbf{F}$ along $\mathbf{D}$.

   (f) Sketch a diagram illustrating the three vectors of part (e), as well as the component of $\mathbf{F}$ that is perpendicular to $\mathbf{D}$.

9. (a) Let T be the transformation given by

   $$A = \begin{bmatrix} 1 & -1 \\ 2 & 0 \end{bmatrix}$$

   Show how T transforms the unit square $(0 \leq x \leq 1, 0 \leq y \leq 1)$.

   (b) Let R be the transformation given by rotation of the plane through an angle $\pi/4$ clockwise. Find the matrix for R. (Hint: convert polar coordinates to Cartesian.)

10. Suppose $f(-1) = 1$, $f'(-1) = -2$, $f''(-1) = 0$, $f'''(-1) = 6$, and $f^{(4)}(-1) = 3$. Write a polynomial that has the same value as $f(x)$ at $x = -1$, and that has the same first four derivatives as $f(x)$ at $x = -1$.

11. Give the Taylor series centered at $x = 0$. Write the first four nonzero terms and indicate the form of the general term.

 (a) $1/(1 + x/2)$

 (b) The "hyperbolic cosine" function: $\cosh x = \frac{1}{2}(e^x + e^{-x})$.

12. (a) Sketch the polar graph $r = 2\sin\theta$ for $0 \leq \theta \leq \pi$.

 (b) Sketch the polar graph $r = \theta$ for $0 \leq \theta \leq 2\pi$.

13. Let $f(x) = \begin{cases} \dfrac{1-\cos x}{x^2} & \text{for } x \neq 0 \\ A & \text{for } x = 0 \end{cases}$

 (a) To make f continuous, we define A to be $\displaystyle\lim_{x\to 0} \frac{1-\cos x}{x^2}$. Compute this value.

 (b) Use series to approximate

$$\int_0^1 f(x)dx = \int_0^1 \frac{1-\cos x}{x^2}\,dx$$

 with an error no more than $0.001 = 1/1000$. Roughly how large is the error?

14. All you know about a function f is that $f(8, 5) = 33$, $f_x(8, 5) = 3$ and $f_y(8, 5) = -1$.

 (a) Estimate the value $f(8.01, 5.02)$.

 (b) Give the equation of the tangent plane to the graph of f at $(8, 5)$ in intercept form.

15. Determine if the series converges or diverges. Explain your reasoning. Give the sum S exactly if possible, or else determine how large n has to be so that the partial sum S_n approximates S with an error less than $1/100$.

 (a) $\displaystyle\sum_{j=3}^{\infty} \frac{2^{j+1}}{7^j}$ (b) $\displaystyle\sum_{n=1}^{\infty} (-1)^n \frac{1}{n^3}$

16. The density function for the life span t (in months, $t \geq 0$) of a randomly selected transistor is $p(t) = Ae^{-At}$. We find that 10% of all transistors fail in the first six months.

 (a) What is the value of A to four decimal places?

 (b) What is the probability that a randomly selected transistor lasts 3 years or longer?

17. (a) Match up the hidden line 3-D diagram with the corresponding contour diagram.

 (b) For each contour plot, give the coordinates of any saddle points, or clearly mark the locations.

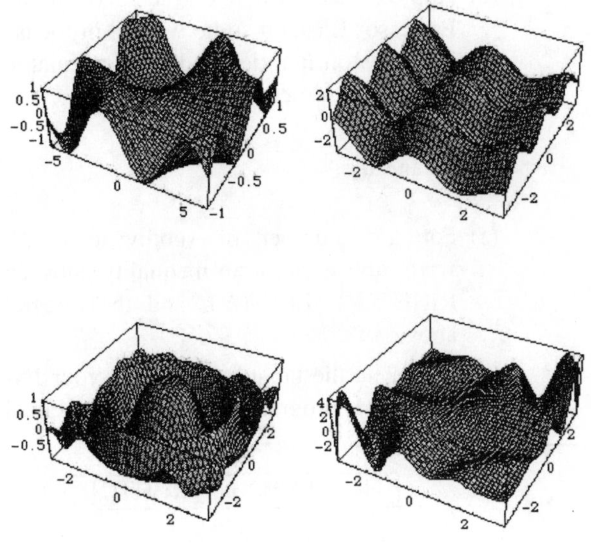

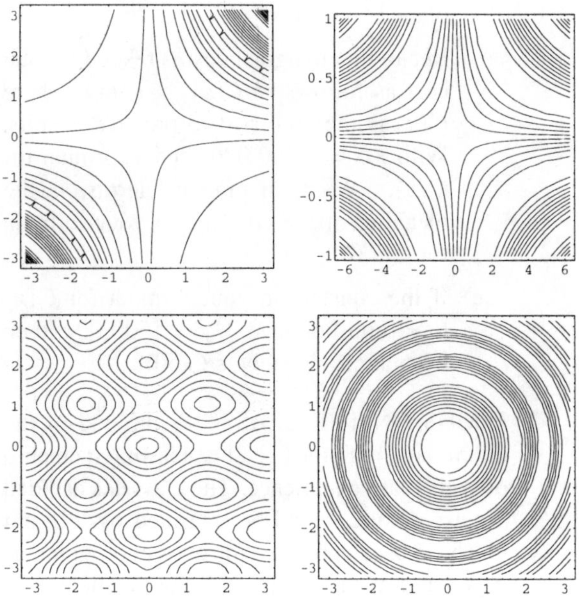

Final Examination for Calculus 2: Test 2

Take Home Portion:

DIRECTIONS: You may use your text, notes, and calculator. Be sure all solutions are clearly written and that your methods are explained. It is recommended you work out all problems in rough draft form before writing them in final form on this exam paper. The accuracy of approximations should be discussed.

1. A "PLANE" PROBLEM: A study of the costs to produce airplanes in World War II led to the theory of "learning curves," the idea is that the marginal cost per plane decreases over the duration of a production run. In other words, with experience, staff on an assembly line can produce planes with greater efficiency. The 90% learning curve describes a typical situation where the marginal cost, MC to produce the xth plane is given by $MC(x) = M_0 x^{\log_2 0.9}$, where M_0 is the marginal cost to build the first plane.

   (a) If a plant produces planes with a 90% learning curve on production costs, and the marginal cost for the first plane is \$500 thousand, then what is the marginal cost to produce the second plane? The fourth plane?

   (b) Recall that marginal cost is related to total cost as follows: $MC(x) \approx C'(x)$, where $C(x)$ is the total cost to produce x units. With $M_0 =$ \$500,000, find a formula for $C(x)$. What, in practical terms, is the meaning of the constant in your formula for $C(x)$?

   (c) If the constant in your formula for $C(x)$ is \$20 million, and $M_0 =$ \$500,000, then what approximately is $C(50)$?

2. FOCUS ON THE CLASS: It's the end of the quarter and it's time to gather a lasting memento of your Math experience. The class has decided a class picture would be treasured by all, and you have been appointed the class photographer! You stand at the origin with your camera, and your classmates, along with your instructor, are strung out along the curve $y = 1/e^x$ from $(0, 1)$ to $(2, 1/e^2)$.

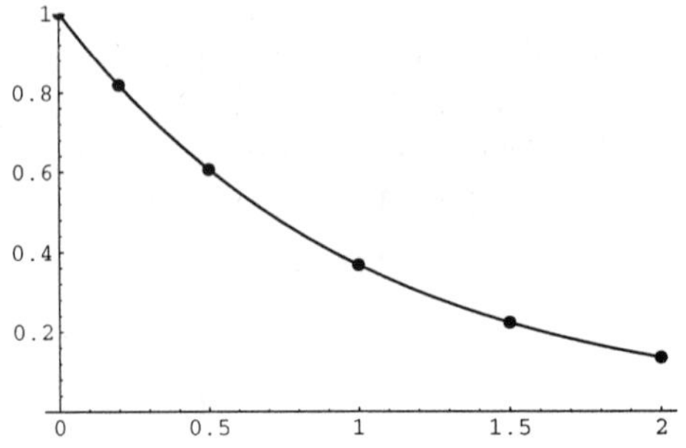

   (a) As a function of x, what is your distance to your classmate at a point $(x, 1/e^x)$ on the curve?

   (b) Write an integral that gives the average value of the distance function in (a).

   (c) Use your calculator to evaluate the integral in (b) to 4 decimal places. Explain your method and justify your accuracy.

   (d) You focus your camera according to part (b). Who is more in focus, SUSAN at $x = 0.56$ or PHIL at $x = 1.39$? Justify your answer.

   (e) Approximately where on the curve should Professor E. stand so he will be in focus? Use Newton Iteration and your calculator. Explain your method.

3. $\displaystyle\int_{-10}^{16} \arctan x \, dx$

   (a) For any number of subdivisions, N, **write and explain** an inequality between RIGHT(N), LEFT(N), and the integral above.

   (b) Complete the table below using your **INTEGRAL** program. (Round to 4 decimal places.)

N	LEFT(N)	RIGHT(N)
2		
3		
4		

(c) Use the graph of $\arctan x$ to explain the wide variations in the table above. Use a separate graph for $N = 2$, $N = 3$, and $N = 4$.

(d) Find a so that you can get overestimates and underestimates for the integrals on the right via the midpoint and trapezoid rules:

$$\int_{-10}^{16} \arctan x\,dx = \int_{-10}^{a} \arctan x\,dx + \int_{a}^{16} \arctan x\,dx.$$

(e) Find MID(50) and TRAP(50), accurate to 4 decimal places, for the two integrals on the right in (d), and use these results to find numbers A and B such that:

$$A \le \int_{-10}^{16} \arctan x\,dx \le B.$$

Explain.

(f) Use integration by parts and the Fundamental Theorem of Calculus to evaluate the original integral and compare with the results in (e).

4. OPTIMIZATION: Do one of the following optimization problems. You may do the other problem for extra credit .

(a) A piece of wire 100 cm long is to be cut into several pieces and used to construct the skeleton of a box with a square base.

i. What is the largest possible volume that such a box can have?

ii. What is the largest possible surface area?

(b) You have decided to build the Erickson Coliseum as a tribute to your favorite instructor. Your contractor estimates the initial costs (buying the island of Manhattan upon which to build) as 400 times the cost of the first floor. The second floor is projected to cost twice as much as the first floor, the third floor three times as much as the first floor, etc. What number of floors in the building will give the cheapest cost per floor?
NOTE: $1 + 2 + 3 + \cdots + n = n(n+1)/2$.

In-Class Portion: Part I

DIRECTIONS: You may use your notes, course supplement, and calculator, but not your text. Be sure it is clear how solutions were determined.

1. Find each of the following indefinite integrals by the method indicated:

(a) $\displaystyle\int \frac{e^x}{\sqrt{3e^x+1}}\,dx$ *(substitution)*

(b) $\displaystyle\int (x \sin x)\,dx$ *(integration by parts)*

2. Evaluate $\displaystyle\int_1^{e^5} \frac{3}{x}\,dx$ using the Fundamental Theorem of Calculus.

3. Suppose $p(t)$ gives the rate at which telephone calls come into a regional switching center at time t hours where t goes from 0 (midnight) to 24. What does $\int_9^{14} p(t)\,dt$ mean in practical terms?

4. A manufacturer finds that the marginal cost of producing x units of a certain product is $C'(x) = 2x + 5x^{3/2}$. Find the cost function, $C(x)$ if $C(1) = 8$.

5. Find the value of each definite integral using the graph of $f(x)$ given below.

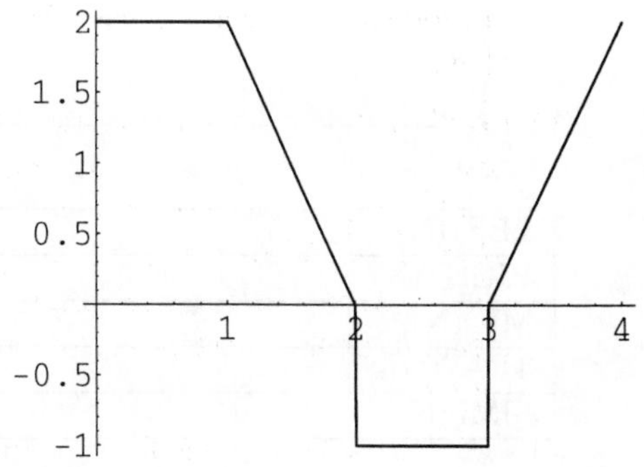

(a) $\displaystyle\int_0^4 f(x)\,dx$

(b) $\displaystyle\int_1^1 f(x)\,dx$

(c) $\displaystyle\int_0^3 |f(x)|\,dx$

6.

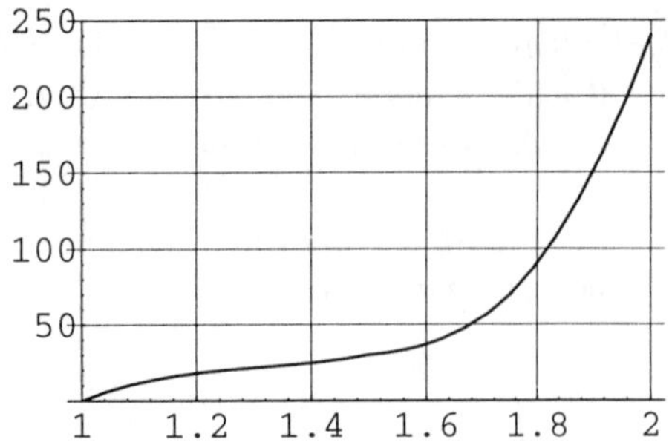

Let g be the function shown graphically. When asked to estimate $\int_1^2 g(x)\,dx$, a group of students submitted the following answers: -4, 4, 45, and 450. Only one of these responses is reasonable; the others are "obviously" incorrect. Which is the reasonable one? Why?

7. The following numbers were obtained by approximating $\int_0^1 f(x)\,dx$: 0.36735, 0.39896, 0.36814, 0.33575. Each approximation used the same number of subdivisions. Place the numbers in the appropriate place in the table and find **SIMP(N)**.

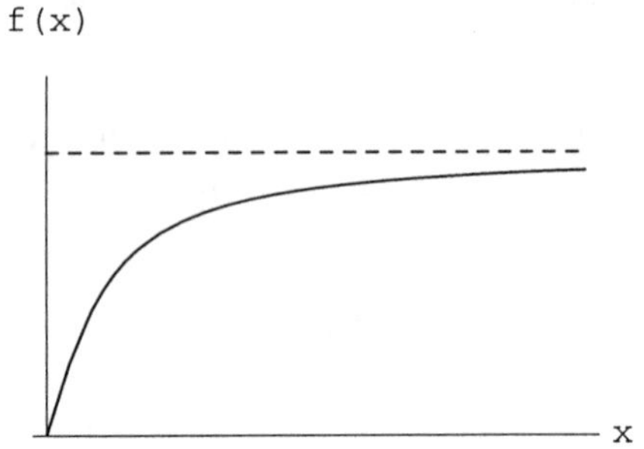

LEFT(N)	
RIGHT(N)	
MID(N)	
TRAP(N)	
SIMP(N)	

8. The graph of $f'(x)$ is given below.

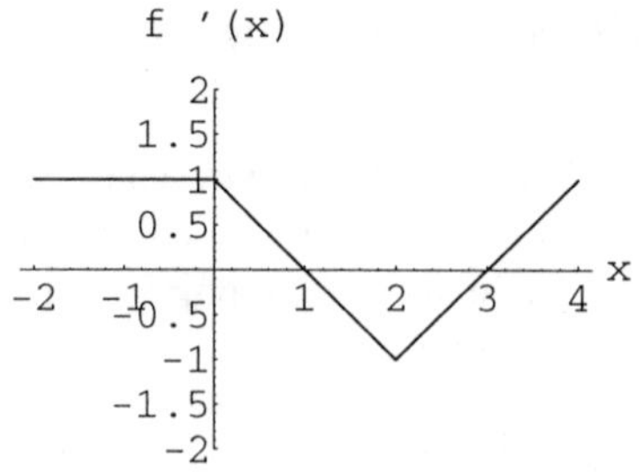

(a) On the same axes, sketch and label the graph of $f(x)$ such that $f(0) = 0$.

(b) From the graph, estimate $\int_0^3 f'(x)\,dx$.

(c) State the values of x at which $f(x)$ has: local maxima, local minima, points of inflection.

9. A mouse moves back and forth in a tunnel, attracted to bits of cheese alternately introduced to and removed from the right and left ends of the tunnel. The graph of the mouse's velocity is given at right. Assume that the mouse starts ($t = 0$) at the center of the tunnel. Complete the table below.

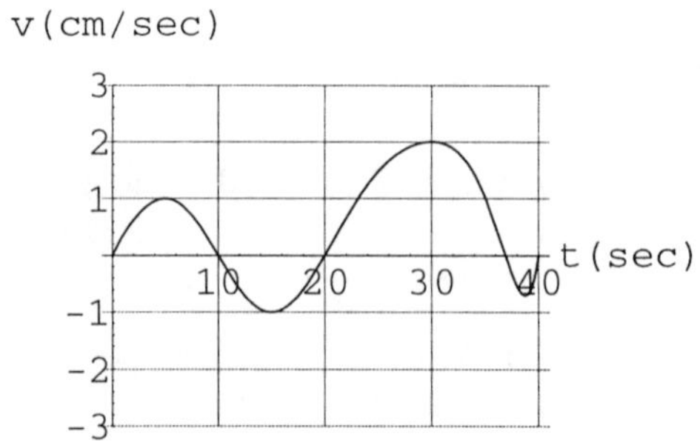

Use the graph to estimate the times at which:

(a) The mouse changes direction.

(b) The mouse is moving most rapidly to the right.

(c) The mouse is farthest to the right of center.

(d) The mouse's **speed** is decreasing.

(e) The mouse is at the center of the tunnel.

In-Class Portion: Part II

DIRECTIONS: Do one of the following two problems presented in class as part of Project 2.

1. Tickets to the "Cal Culus and the Derivatives" concert go on sale the day of the performance. To get seats, calculus fans start arriving at the concert hall at 8 o'clock. At right is the graph of their arrival rate r (people/hr) over the next few hours. Suppose the doors open at 9 o'clock, and, from that time on, concert-goers are allowed in at the rate of 100 per hour. **Your job is to interpret the following quantities using the graph.**

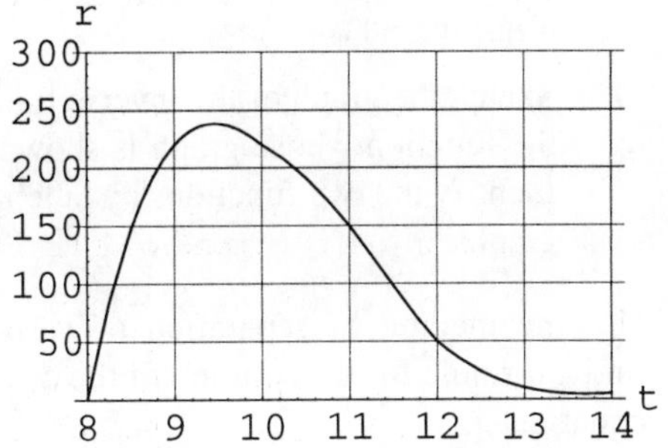

   (a) The length of time a person who arrives at 10 o'clock has to stand in line.

   (b) The rate at which the line is growing at 11 o'clock.

   (c) The time at which the length of the line is a maximum.

2. Throughout much of this century, the yearly consumption of electricity in the U.S. has been increasing exponentially at a continuous rate of 6%. Assuming this trend continues, and that the electrical energy consumed in 1900 was 1.6 million megawatt-hours,

   (a) Find the average yearly electrical consumption since 1920.

   (b) Find the year in which electrical consumption was closest to the average obtained in (a).

   (c) What is the rate of change of consumption when t is the year determined in (b)?

Final Examination for Calculus 2: Test 3

1. Evaluate each indefinite integral, then check your answer by differentiating it.

   (a) $\int 2x^2(x^3+4)^5\,dx$

   (b) $\int x\cos(2x)\,dx$

2. (a) Find the *signed area* of the shaded region.

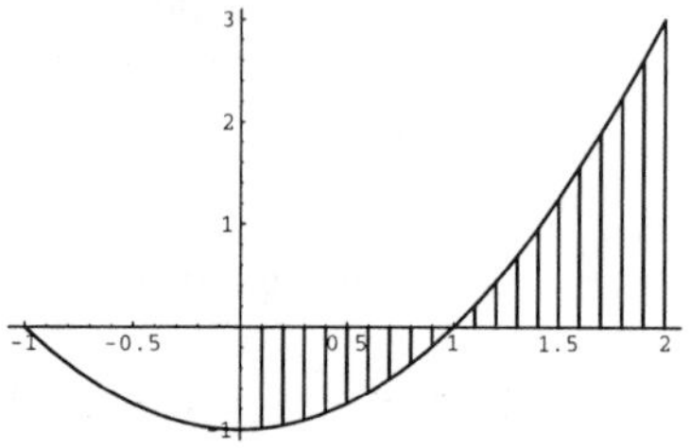

   (b) Approximate the signed area of the region shown above by using a left endpoint Riemann sum with subintervals of width $\Delta x = \frac{1}{2}$.

3. Evaluate each integral. Use correct mathematical notation.

   (a) $\int_0^{\ln 2} e^{3x}\,dx$

   (b) $\int_2^X 3x^2\,dx$

   (c) $\int_1^\infty \frac{1}{x^3}\,dx$

4. (a) Use the right triangle shown below to find a formula for $\tan(\alpha)$ in terms of x and to find a formula for β in terms of x.

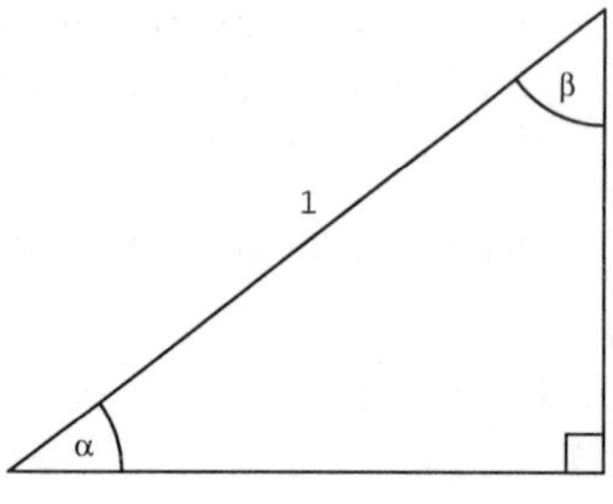

   (b) Evaluate $\frac{d}{dx}\big(\arctan(x)\big)$.

   (c) Evaluate $\int \frac{1}{1+x^2}\,dx$.

   (d) Evaluate $\arccos(\sqrt{3}/2)$.

   NOTE: Your answer to (d) may contain neither degrees nor decimals! Give your answer in radians (for full credit).

   (e) Sketch the graph of the inverse function of the function whose graph is shown below. Identify the two functions by labeling their graphs.

5. Use the method of separation of variables to find a formula for the solution of the differential equation

$$\frac{dy}{dt} = -3xy^2.$$

   Your answer should include an arbitrary constant. *Then* verify that your solution is correct by showing that it satisfies the differential equation.

6. Use the formula

$$\int \frac{dx}{\sqrt{x^2 \pm a^2}} = \ln\left|x + \sqrt{x^2 \pm a^2}\right| + C$$

   to find a formula for the indefinite integral

$$\int \frac{dx}{\sqrt{x^2+6x+13}}.$$

7. (a) Find the amplitude, period, and frequency of $f(x) = 2\sin(3x)$, then sketch $y = f(x)$ on the axes below. Your graph should show clearly the amplitude and period of $f(x)$.

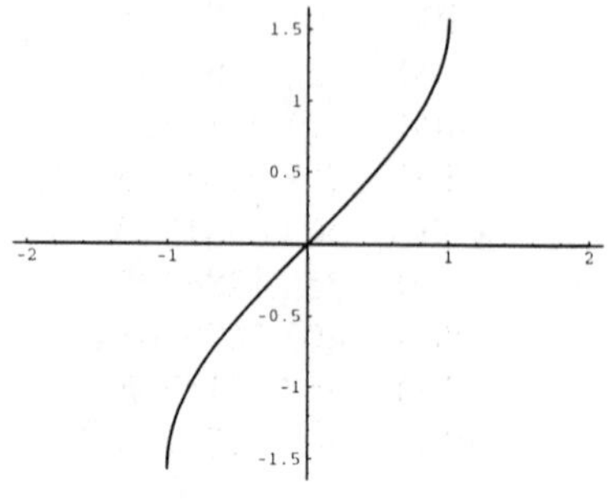

(b) The graph of a periodic function $y(t)$ is shown below. Use it to estimate the amplitude, period, and frequency of $y(t)$.

8. (a) Use the formula for the nth Maclaurin polynomial for a function $f(x)$ given below to find the 5th Maclaurin polynomial for $f(x) = \ln(1+x)$. Show all of your work.

$$p_n(x) = f(0) + f'(0)x + \frac{f''(0)}{2!}x^2 + \frac{f^{(3)}(0)}{3!}x^3 + \cdots + \frac{f^{(n)}(0)}{n!}x^n$$

(b) Why must we use Taylor polynomials (or other approximation techniques) to *approximate* values of such functions as $\sin(x)$, e^x, $\ln(x)$, and $\sqrt[3]{1+x}$, but not to compute values of such functions as $x^3 - 3x + 2$ and $(x-3)/(x^2+1)$?

9. In each part of this problem, partial sums for an infinite series are given. Use these partial sums to determine if the series converges or diverges. If the series converges, give its sum to as many places after the decimal point as you are sure of. If the series diverges, explain why.

(a)

n	*Partial Sum*
1	1.0000000
100	0.8224175
101	0.8225156
200	0.8224546
201	0.8224793
300	0.8224615
301	0.8224725
400	0.8224639
401	0.8224701
500	0.8224650
501	0.8224690

(b)

n	*Partial Sum*
100	51.6421711
200	101.8154571
300	151.9168232
400	201.9887437
500	252.0445296
600	302.0901099
700	352.1286476
800	402.1620305
900	452.1914762
1000	502.2178163

(c)

n	*Partial Sum*
1	0.0909091
1000	0.0053337
1001	0.0886739
2000	0.0053354
2001	0.0886722
3000	0.0053360
3001	0.0886716
4000	0.0053363
4001	0.0886713
5000	0.0053364
5001	0.0886712

10. (a) Write the following series in summation notation, than find its sum.

$$1 - \frac{1}{3} + \frac{1}{9} - \frac{1}{27} + \frac{1}{81} - \cdots$$

(b) Write out the first 4 terms of the series, then find its sum.

$$\sum_{n=0}^{\infty} \frac{2}{3^n}$$

(c) Write the 3rd degree Maclaurin polynomial for the function $1/\sqrt[3]{1+x}$. You are not required to simplify your answer. [HINT: Use your list of Taylor series.]

What value of x would you use in this polynomial in order to approximate $1/\sqrt[3]{1.02}$?

11. (a) Find all values of x for which the power series

$$\sum_{n=1}^{\infty} \frac{x^n}{n}$$

converges and all values of x for which the power series diverges by applying appropriate convergence/divergence tests. Be sure to show all of your work, to give reasons for all of your answers, and to use correct mathematical notation.

(b) To what function does the power series in part (a) converge when it converges? Explain (briefly).

12. Compute the limits. Show your work!

(a) $\lim\limits_{x \to 0} \dfrac{1-\cos x}{2\sin x}$

(b) $\lim\limits_{x \to 0} \dfrac{1-\cos x}{x^2}$

(c) $\lim\limits_{x \to \infty} \dfrac{e^x}{x^2}$

(d) $\lim\limits_{x \to 0^+} x \ln x$

Final Examination for Calculus 2: Test 4

Certain problems require explanation. These are labeled **Explain**. *For these you should write in complete sentences and include pictures and/or formulas if you think it helps the explanation. Problems without* **Explain** *do not require explanations, but your solutions should be neatly and clearly presented so we understand what's going on. Remember to include units in your answers whenever appropriate.*

1. It is estimated that during the next year the population of a certain town will be changing at the rate of $4 + t^{2/3}$ people per month, where t represents time in months starting now. If the current population is 10,000, what will be the population 8 months from now?

2. Given that $\int_{-1}^{0} f(x)\,dx = 3$, $\int_0^1 f(x)\,dx = -1$, and $\int_{-1}^{1} g(x)\,dx = 7$, find the value of $\int_{-1}^{1}(3g(x) + 2f(x))\,dx$.

3. Find the value or values of x for which $(1 + x + x^2 + x^3 + \ldots) = 5$.

4. The graphs of two functions which we call $g(x)$ and $h(x)$ are shown below. The function $g(x)$ is an antiderivative of $h(x)$.

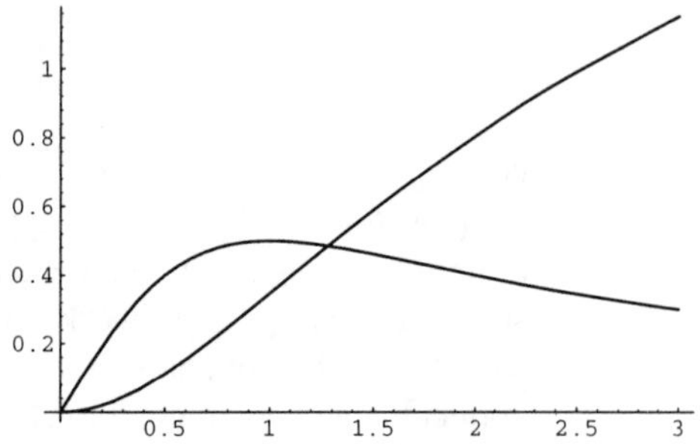

   (a) Label which graph is that of $g(x)$ and which is that of $h(x)$.

   (b) Only one of the following four equations is correct. Circle the correct one.

$$g(x) = \int h(x)\,dx \qquad g(x) = h'dx$$

$$g(x) = \int_0^x h(t)\,dt \qquad g(x) = h(x) + C$$

5. The table below lists the exact values at $x = 1$ and $x = 1.5$ for a function $f(x)$, an antiderivative $F(x)$ of $f(x)$, and the first two derivatives $f'(x)$ and $f''(x)$ of $f(x)$.

x	$f(x)$	$F(x)$	$f'(x)$	$f''(x)$
1	2	3.1	0.3	1.2
1.5	2.6	2.5	−0.4	0.2

   (a) Which constant function best approximates $f(x)$ near $x = 1$?

   (b) Which quadratic function best approximates $f(x)$ near $x = 1$?

   (c) Find the exact value of $\int_1^{1.5} f(x)\,dx$.

   (d) Based on your answer to question (5c), can you tell whether there exists a value of x between 1 and 1.5 for which $f(x) < 0$? Briefly explain your reasoning.

6. A function $p(x)$ satisfies $p(0) = 1$ and the graph of its derivative $p'(x)$ is shown below. Circle which of the four functions is the degree two Taylor polynomial for $p(x)$ centered around 0.

$$1 + x + x^2 \qquad 1 + x - x^2$$

$$1 - x + x^2 \qquad 1 - x - x^2$$

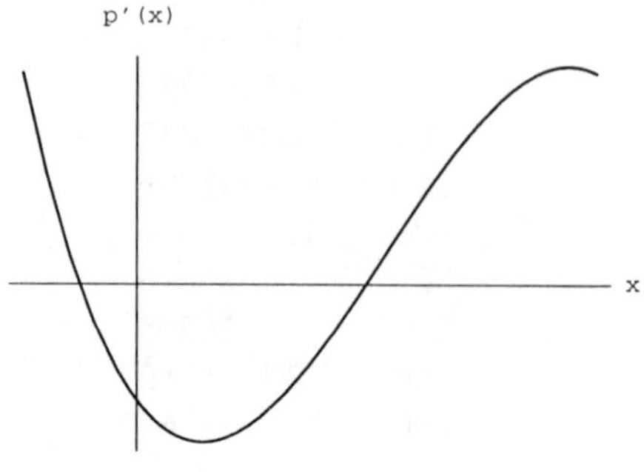

7. For certain constants a and b the population of bacteria in a dish satisfies the differential equation $dP/dt = aP - bP^2$, where P is the population in millions of bacteria and t is time in hours. At time $t = 0$, the population is 10 million and growing at the instantaneous rate of 2 million bacteria per hour. After a long time, the population stabilizes at 50 million. (The following questions should be answered without solving the differential equation.)

   (a) Sketch a graph of P as a function of t. Be sure to include all intercepts and asymptotes, if any.

   (b) Find the constants a and b.

   (c) Suppose that when t is 500, you add 75 million bacteria from another source to the dish. Will the population still stabilize at 50 million, will it stabilize at another number, or is it impossible to determine what happens to the population under these circumstances? Explain.

8. An object is moving in the plane and its position is given by the parametric equations:

$$x = 1 + 2\cos t$$
$$y = -\sin t$$

for $0 \leq t \leq \pi$.

   (a) Show that the object always is moving along the ellipse $(x-1)^2/4 + y^2 = 1$.

   (b) The outline of this ellipse is shown below.

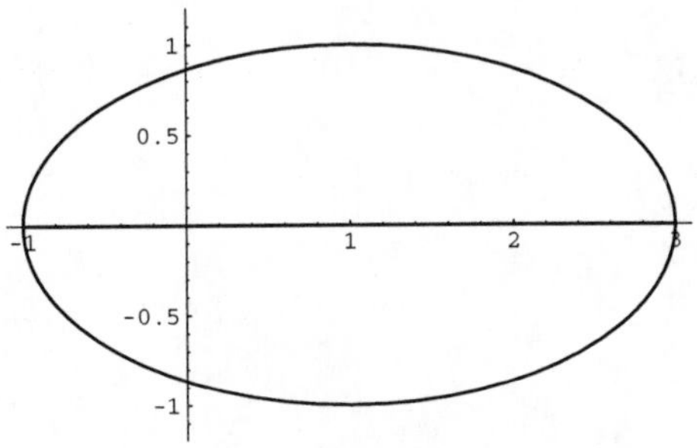

   Mark the points on this ellipse corresponding to the times $t = 0$ and $t = \pi$, and draw arrows to indicate the direction of the object as it travels between these two points.

   (c) Notice that the point $\left(\frac{1}{2}, -\frac{\sqrt{3}}{2}\right)$ is on the ellipse. Indicate its location on the ellipse. What is the slope of the tangent line to the ellipse at this point?

9. Suppose that starting January 1, 1995 you receive income at a continuous rate of $100/(1+t)$ dollars per year, over a ten-year period. Assume that the interest rate is 10% compounded continuously and $t = 0$ represents January 1, 1995.

   (a) Write an integral which represents the present value of this income stream.

   (b) Let LEFT(n) be the approximation to the value of this integral obtained using a left hand sum with n subdivisions. Is LEFT(10) an under-estimate or an over-estimate of the integral? State the *single* property of the function being integrated which justifies your answer.

   (c) Approximate the present value of this income stream to the nearest dollar. You may use any method you like so long as you justify how you know for certain that your answer is correct to the nearest dollar. Explain.

10. A company manufactures light bulbs, and the graph of the density function $p(t)$ which measures the probability that a given light bulb will last for more than t hours is shown below. Every bulb lasts for at least 200 hours and the maximum time a bulb will last is 1000 hours.

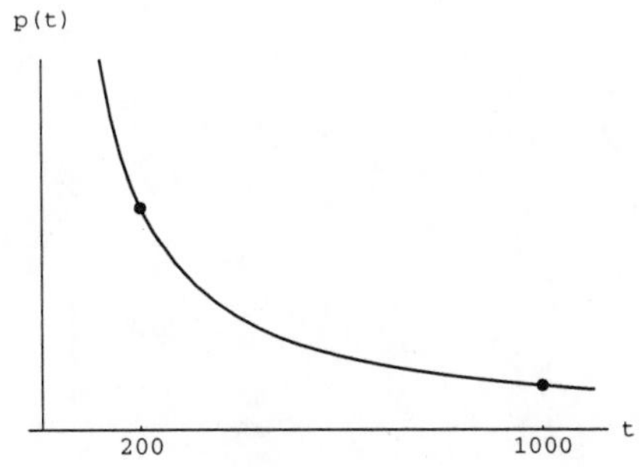

   (a) Sketch the corresponding cumulative density function $C(t)$.

   (b) Explain in one complete sentence what $C(600) - C(400)$ represents in practical terms.

   (c) Express the number $C(600) - C(400)$ in terms of the function $p(t)$.

11. The phase plane for the system of differential equations $dx/dt = -0.05y$, and $dy/dt = -0.01x$ is shown below.

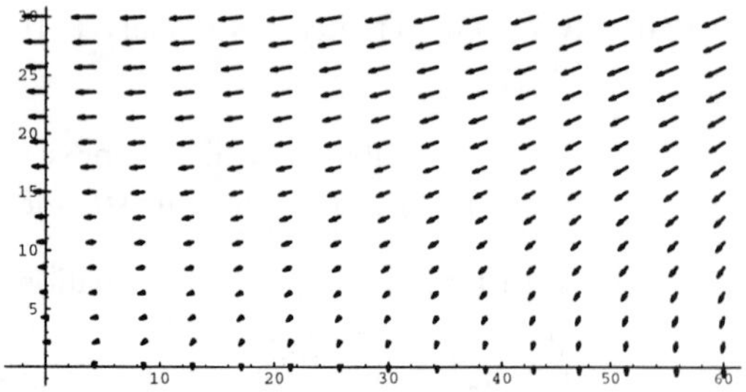

(a) Draw the trajectory which goes through the point (50,15). Be sure to draw arrows on the trajectory to indicate its direction and briefly explain how you determined its direction.

(b) If we divide dy/dt by dx/dt we obtain the differential equation $dy/dx = x/5y$. Use the method of separation of variables to solve this equation and find the equation of the trajectory you drew in part (a).

Final Examination for Calculus 2: Test 5

1. Discuss what each of the following represents for a nonnegative function f defined on $[a, b]$.

   (a) Δx_k

   (b) c_k

   (c) $f(c_k)\Delta x_k$

   (d) $\sum_{k=1}^{n} f(c_k)\Delta x_k$

   (e) $\lim_{n\to\infty} \sum_{k=1}^{n} f(c_k)\,\Delta x_k$

   (f) $\int_a^b f(x)\,dx$

2. In the development of the formula for finding arc length, we made use of the knowledge that a polygonal path can be used to approximate the length of a curve. Discuss, without giving all the nitty-gritty details but by referring to an arbitrary partition, how this was used. This includes mentioning what $\sqrt{(\Delta x_k)^2 + (\Delta y_k)^2}$ represents.

3. Explain why, in defining the inverse of $\cosh x$, you have to restrict the domain of $\cosh x$, and why, in defining the inverse of $\tanh x$, you don't have to restrict the domain of $\tanh x$.

4. Using knowledge about the graphs of a function and its inverse, graph $\cosh^{-1} x$. What knowledge are you using?

5. Give three different types of functions (e.g., logarithmic, trigonometric, algebraic) whose derivatives are rational functions.

6. Using the definition of $\sinh x$, argue that $\sinh x$ approaches $e^x/2$ as x increases without bound.

7. Use the definition of the convergence or divergence of an improper integral to decide convergence or divergence of

$$\int_0^\infty \frac{dx}{(x+1)^2}.$$

8. Is $\int_0^1 \frac{dx}{x}$ convergent or divergent? Explain.

9. Give an example of a sequence that is

   (a) Increasing and divergent.

   (b) Increasing and convergent.

   (c) Decreasing and convergent.

   (d) Decreasing and divergent.

10. Define what it means for a series $\sum a_n$ to converge.

11. State the nth term test for divergence of a series.

12. Argue that the sequence of partial sums $\{ s_n \}$ of a series $\sum a_n$ of nonnegative terms is a nondecreasing sequence.

13. For what reason(s) do we want to know whether a series converges absolutely?

14. Knowing that

$$\sin x = x - \frac{x^3}{3!} + \frac{x^5}{5!} - \frac{x^7}{7!} + \cdots,$$

express $\int \sin x^2\,dx$ as a power series.

15. What is the major knowledge that you gained from the unit on power series? Include in your response reasons for having power series (that is, what do they do for us).

16. Find the Taylor polynomial of order 3 generated by $f(x) = \sqrt{x}$ at $a = 4$.

17. Find the radius and interval of convergence of $\sum \frac{x^n}{\sqrt{n^2+3}}$.

18. Prove that the derivative of $\sinh^{-1} x$ is $(1+x^2)^{-1/2}$. (Hint: use the definition of $\sinh^{-1} x$ and implicit differentiation.)

19. Let R be the region to the right of the y-axis, below $y = e^{-x}$, and above the x-axis. Find, if possible, the

   (a) Area of R.

   (b) Volume of the solid obtained by revolving R about the y-axis. (Just set up; don't evaluate.)

20. Find the volume of the solid of revolution obtained by revolving about the y-axis, the region bounded by $y = e^x$, the x-axis, and $x = 1$ and $x = 2$, using the *shell* method. (Just set-up the integral. Don't evaluate it.)

21. Discuss *in general* how the integral formula is derived for finding the volume of a solid of revolution via disks. A drawing should accompany your explanation. Begin with a region to revolve, construct a partition, etc.

22. Evaluate these integrals:

(a) $\int_0^1 \sqrt{(4-x^2)}\,dx$ (c) $\int \frac{(1-x)^2}{x^{1/3}}\,dx$

(b) $\int \frac{dx}{x^3+x}$ (d) $\int \tan^4 x\,dx$

23. Graph $r^2 = 4\sin 2\theta$. (You may use your calculator.)

24. Find the points of intersection of $r = 3$ and $r = 3\cos 2\theta$. (You can verify your work with a graph, but I want to see how you do it analytically.)

25. Show that the point $(2, 3\pi/4)$ lies on the curve $r = 2\sin 2\theta$.

26. Find the area of the region that lies inside the circle $r = 1$ and outside the cardiod $r = 1 - \cos\theta$. (Just set up. Don't evaluate.)

27. Mention two reasons why it can be "tricky" working with polar coordinates. That is, mention two areas of study in polar coordinates where one has to be cautious, and why.

28. Find the area within the first turn of the spiral $r = e^\theta$ for $0 \le \theta \le 2\pi$.

29. A curve is given parametrically as $x = \tan y$, $y = \sec t$.

(a) Eliminate the parameter t to find an equation in x and y for the curve.

(b) Graph the curve.

(c) Use parametric equations to find the slope of the curve at the point where $t = \pi/4$.

(d) Find all values of t where the slope of the graph at $\big(x(t), y(t)\big)$ is $1/2$.

30. Find parametric equations for the circle $x^2 + y^2 = a^2$, using as parameter the arc length s measured counterclockwise from the point $(a, 0)$ to the point (x, y).

31. Set up, but don't evaluate, an integral that gives the length of the curve $x = 5t^2$, $y = t^{1/2}$, from point $(0, 0)$ to the point $(5, 1)$.

32. Describe how to find, geometrically, (a) the difference of two vectors $(\mathbf{v}_1 - \mathbf{v}_2)$, and (b) a scalar multiple of a vector $(a\mathbf{v}_1)$. (I want words here, not drawings.)

33. What important role do the vectors **i**, **j**, and **k** play? That is, what is our major use of them?

34. Define, geometrically, the dot product of two vectors.

35. (a) Find the scalar component of $\mathbf{X} = -2\,\mathbf{i} + 3\,\mathbf{j}$ in the direction of the vector $\mathbf{Y} = 4\,\mathbf{i} - 5\,\mathbf{j}$.

(b) Now find the vector component of **X** in the direction of **Y** (call it **Z**).

(c) Graph **X** and **Y** so that their tails are at the origin.

(d) Use what you did in (c) to draw the vector component of **X** in the direction of **Y**.

(e) Also, use what you did in (c) to draw a vector **W** normal to **Y** such that **X** is resolved in **Z** and **W**.

(f) Resolve **W** into **i** and **j**.

36. Let $\mathbf{A} = 5\,\mathbf{i} - \mathbf{j} + \mathbf{k}$, $\mathbf{B} = 5\,\mathbf{j} + 5\,\mathbf{k}$, and $\mathbf{C} = -15\,\mathbf{i} + 3\,\mathbf{j} - 3\,\mathbf{k}$. Which pairs of vectors, if any, are (a) parallel? (b) normal? Explain why they are parallel and why they are normal.

37. Find two unit vectors normal to both $\mathbf{A} = 2\,\mathbf{i} + \mathbf{j} - \mathbf{k}$ and $\mathbf{B} = \mathbf{i} - \mathbf{j} + 2\,\mathbf{k}$.

38. Give the geometric interpretation of $\mathbf{A} \times \mathbf{B}$. Explain why $\mathbf{A} \times \mathbf{B}$ gives the area of a parallelogram with adjacent sides of length $|\mathbf{A}|$ and $|\mathbf{B}|$.

39. Find a parametrization of the line segment joining the points $(1, -1, -2)$ and $(0, 2, 1)$.

40. Find an equation for the plane through $A = (1, -2, 1)$ perpendicular to the vector from the origin to A.

41. Find the point in which the line $x = 2$, $y = 3 + 2t$, $z = -2 - 2t$, meets the plane $6x + 3y - 4z = -12$.

Final Examination for Calculus 2: Test 6

End-of-term Essay and Portfolio

*The Final Examination for this course will have two parts, a take-home part and a part given as a "regular" exam at the scheduled time. The take-home part, on which you may (indeed, should) start now, consists of an essay on the topic "***What's Important About Approximation of Functions by Taylor Polynomials***."*

You may talk to anyone and use any materials as you prepare your essay. You must acknowledge any sources used (people as well as written materials), and you must write your own essay in your own words. Use EXP for typing and printing your essay, which must not be longer than three (3) pages (with default type size, line spacing, and margins). Your essay counts for one-third of your final exam grade.

Your portfolio is due at the start of the scheduled exam. The portfolio should consist of the following:

1. *Your end-of-term essay.*
2. *Your individual report for Lab 22.*
3. *One other double-submission written report.*
4. *One more sample of your work during the term (lab or project report).*

You may submit copies (not necessarily originals) of your work. Place all four items in a manila folder with your name on it.

In-Class Portion

One-third of your final exam grade will come from your end-of-term essay, the other two-thirds from this exam. The exam has four problems; in order, their weights are 5%, 25%, 30%, 40%. We suggest that you allocate up to two hours, in more or less the same proportions, for working out these problems. Then use your remaining time for checking, refinement, correction, or finishing anything you didn't complete in the first pass.

If you find that a calculation leads to an unreasonable answer, you will get more credit for identifying the apparent problem and saying why the answer is unreasonable, less credit for not noticing or for pretending there is no problem. Of course, you will get still more credit if you find what went wrong in the calculation and fix it.

You may use your text, project, and lab materials, your tables, your calculator, and your notes. Show your work for each problem.

1. Evaluate $\sum_{k=2}^{\infty}\left(\frac{1}{5}\right)^k$.

2. Consider a four-sided pyramid with a square base; the x-axis is a line of symmetry. The pyramid is 2 units high, and the square base has sides 1 unit long. The volume of this pyramid is $2/3$ of a cubic unit. Assuming that the pyramid has uniform density, find the x-coordinate of its center of mass.

3. The island nation of Upton in the Indian Ocean has a population of 2,500,000 and a growth rate similar to the rest of the world. The country is ruled by a military junta that allows no immigration or emigration. The growth of Upton's population may be modeled by the initial value problem

$$\frac{dP}{dt} = kP^2, \quad P(0) = 2{,}500{,}000,$$

where $k = 10^{-8}$, $t = 0$ means now, and time is measured in years.

(a) Show that this model predicts a population explosion in Upton in 40 years.

(b) Unaware of their impending population crisis, the junta plans to permit a net immigration to their island paradise of exactly 10,000 people per year. Modify the differential equation to account for this constant net immigration.

(c) Find a solution of the modified differential equation that satisfies the given initial condition.

(d) When does the modified model predict an infinite population?

4. (a) Use Taylor series or Taylor polynomials to approximate the value of the integral

$$\int_0^{1/2} e^{-2t^2}\, dt$$

to within 0.01. Explain how you know your answer is within 0.01 of the exact value of the integral.

(b) Use an essentially different method to approximate the value of the integral in part (a) to within 0.01.

Final Examination for Calculus 2: Test 7

1. (a) What is the order of contact between e^x and $1+x$ at $x=0$?

   (b) What is the order of contact between e^{-x} and $1-x+\frac{x^2}{2}-\frac{x^3}{3!}$ at $x=0$?

   (c) What is the order of contact between e^{-x} and $1-x+\frac{x^2}{2}+\frac{x^3}{3!}$ at $x=0$?

   (d) What is the order of contact between and $\sin(x)$ and $x-\frac{x^3}{3!}$ at $x=0$?

2. (a) If $f(x)$ and $g(x)$ have order of contact 5 at $x=a$, $F(x)=\int_t^a f(t)\,dt$, and $G(x)=\int_t^a g(t)\,dt$, then what order of contact do you expect $F(x)$ and $G(x)$ to have at $x=a$?

   (b) What order of contact do you expect $f'(x)$ and $g'(x)$ to have at $x=a$?

3. Set a constant c so that the plots of $\sin 3x$ and $1-e^{cx}$ share lots of ink on small intervals centered at 0.

4. What good is the idea of order of contact?

5. Write down the expansion of $1/(1-x)$ in powers of x, and use it to write down the expansion of $1/(1+x^2)$ in powers of x.

6. Write down the expansion of e^x in powers of x, and use it to write down the expansion of e^{-x} in powers of x.

7. Write down the expansion of $\sin x$ in powers of x, and use it to write down the expansion of $\sin 2x$ in powers of x.

8. Write down the expansion of $\cos x$ in powers of x, and use it to write down the expansion of $\cos \pi x$ in powers of x.

9. If all you have is a cheap calculator that only adds, subtracts, multiplies, and divides, then what numbers would you enter to calculate a reasonably accurate numerical estimate of $e^{0.5}$?

10. Use expansions in powers of x to calculate the limits:

    (a) $\lim_{x\to 0} \frac{\sin x}{x}$

    (b) $\lim_{x\to 0} \frac{\sin 3x^2}{x^2}$

    (c) $\lim_{x\to 0} \frac{\sin 4x^4}{2x^4}$

    (d) $\lim_{x\to 0} \frac{1+x-e^x}{x^2}$

    (e) $\lim_{x\to 0} \frac{1-x-e^{-x}}{1-\cos x}$

11. The expansion of

$$f(x)=\sin(\tan x)-\tan(\sin x)$$

in powers of x through the x^7 term is $-x^7/30$. The expansion of

$$g(x)=\arcsin(\arctan x)-\arctan(\arcsin x)$$

in powers of x through the x^7 term is also $-x^7/30$. Use what you see to calculate

$$\lim_{x\to 0} \frac{f(x)}{g(x)}.$$

12. What part of the expansion of a function $f(x)$ in powers of x reflects the behavior of the function for x close to 0?

13. Explain the idea behind Newton's method. Include at least one sketch in your explanation.

Final Examination for Calculus 2: Test 8

Part I

On this part of the exam no calculators or computers may be used. This part is worth 56 points; the entire exam has 150 points. For full credit you must show enough work so that your methods and reasoning are clear.

1. Compute the derivative.

   (a) $\dfrac{d}{dr}\arcsin(r^3)$

   (b) $\dfrac{d}{dt}\ln\sqrt[3]{te^t}$

2. Compute each integral.

   (a) $\displaystyle\int\left(\cos(x/4)+\frac{6}{x^3}\right)dx$

   (b) $\displaystyle\int\frac{t}{\sqrt{t^2-16}}\,dt$

   (c) $\displaystyle\int\tan 3z\,dz$

   (d) $\displaystyle\int x^2\ln x\,dx$

   (e) $\displaystyle\int_0^{\infty} r^2e^{-r^3}\,dr$

   (f) $\displaystyle\int\frac{t+14}{t^2+3t-4}\,dt$

3. Compute the average value of the function $f(w)=1/(16+w^2)$ on the interval $0\le w\le 4$.

4. Compute $\displaystyle\lim_{x\to 0}\frac{e^{-x^2}-1+x^2}{x^4}$.

5. Where possible, determine the sum of the series, stating any conditions that are necessary for convergence. Otherwise explain why the series diverges.

   (a) $\displaystyle\sum_{k=0}^{\infty}\frac{3^k}{8^{k+1}}$

   (b) $\displaystyle\sum_{n=4}^{\infty}\frac{2}{n}$

   (c) $\displaystyle\sum_{j=0}^{\infty}pq^{2j}$

Part II

On this part of the exam calculators or Maple may be used. For full credit you must show enough work so that your methods and reasoning are clear. If you use Maple or a calculator, say what exactly you are having the matching do for you. This part has 94 points.

1. Suppose the graph of $\ln\big(g(x)\big)$ is the one shown below. Find the formula for $g(x)$ itself, and sketch its graph.

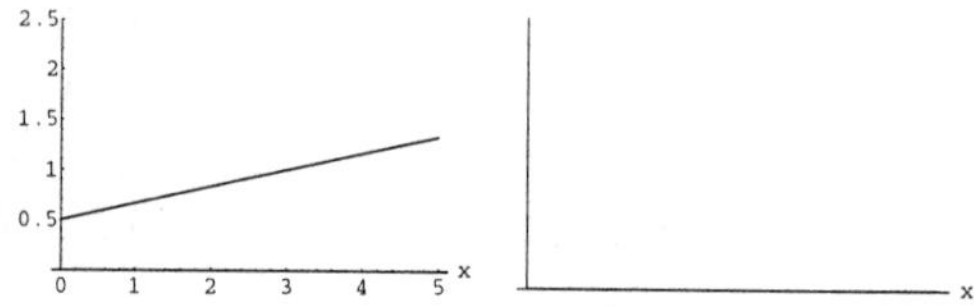

2. Sketch the graph of the function $f(x)=3-x-4/(x-3)^2$. Pay particular attention to the natural domain of f, the critical points and extreme values, and asymptotic behavior.
   NOTE: $f'(x)=-1+8/(x-3)^3$.

3. Suppose $f(-1)=1$, $f'(-1)=-2$, $f''(-1)=0$, $f'''(-1)=6$, and $f^{(4)}(-1)=3$. Write a polynomial that has the same value as $f(x)$ at $x=-1$, and that has the same first four derivatives as $f(x)$ at $x=-1$.

4. Compute the area enclosed between the curves $y=2x^3$ and $y=-x^2+4x+3$.

5. Sketch a graph that is consistent with all the following information:

   - $g(x)$ is defined at all x except for $x=0$
   - $g(-2)=-4$
   - $g'(x)<0$ for $x<-2$ and $x>0$
   - $g'(-2)=0$
   - $g'(x)>0$ for $-2<x<0$
   - $g''(x)<0$ for $x<-3$
   - $g''(x)>0$ for $-3<x<0$ and for $x>0$

6. A particle moves so that $\mathbf{r}(t) = 3\cos t\,\mathbf{i} + \sin t\,\mathbf{j} = (3\cos t, \sin t)$.

   (a) Compute the velocity vector $\mathbf{v}(t)$ and the speed.

   (b) Give an equation in the variables x and y that describes the path of the motion. Sketch the path for $0 \le t \le \pi$, and indicate the direction of motion.

   (c) Compute the length of the path for $0 \le t \le \pi$.

7. Two forces $\mathbf{F} = (2, -5)$, and $\mathbf{G} = (1, 2)$ are acting on a particle.

   (a) What is the combined force $\mathbf{H}$ on the particle?

   (b) Determine the angle between $\mathbf{F}$ and $\mathbf{G}$.

   (c) The particle is displaced from the point $P(-1, 3)$ to the point $Q(3, 0)$. Compute the displacement vector $\mathbf{D}$.

   (d) Compute the work done by the force $\mathbf{F}$ as a particle is displaced along $\mathbf{D}$.

8. We find a Riemann sum approximation $\int_0^{\pi/6} \sec^3 x\,dx \approx 0.60827$ by using the right endpoints of 500 equal size subintervals.

   (a) Does this Riemann sum over- or underestimate the integral?

   (b) Find a good bound E for the error, and find good bounds A and B so that

$$A \le \int_0^{\pi/6} \sec^3 x\,dx \le B.$$

9. Let R be the transformation of the plane given by a counterclockwise rotation through an angle of $5\pi/6$ radians. Find the 2×2 matrix that represents R.

10. Show how $A = \begin{bmatrix} 0 & 2 \\ 1 & -3 \end{bmatrix}$ transforms the standard unit square; label the coordinates at all four corners.

11. A solid is formed by revolving the region bounded by $y = \sqrt{x}e^{-x}$, the x-axis, the y-axis, and $x = 1$ around the x-axis. Set up (but do NOT compute) the integral that gives the volume of this solid.

12. One model describing self-limiting growth is the logistic model. Another, useful in describing the growth of some tumors, is the *Gompertz* model. In this case the rate of change of the quantity y is proportional to y itself, but the constant of proportionality is replaced by a multiplier that declines exponentially over time; i.e., $dy/dt = ke^{-at}y$.

   (a) If $k = 0.3$, $a = 0.6$, and $y(0) = 6$, solve for y as a function of t.

   (b) Describe what happens to y as $t \to \infty$.

13. The local gas station/video rental/convenience store has three staff persons working from opening at 7 am to 2 pm, two working from 2 pm to 4 pm, four working from 4 pm to 10 pm, and one from 10 pm to closing at 1 am. One fine day everyone works as usual, except the two on duty from 2 pm to 4 pm lock up at 3 pm and sneak out to play street hockey on the roof for an hour. Graph the work done (in staff-hours) over the course of this day as a function of time.

14. We are back at the U of I's famous Morrow plots. Two fertilizer treatments A and B are being compared to the unfertilized control. The three different cumulative distribution functions for the heights of the plants at maturity are shown below; A is on the left, B is on the right, and the middle graph represents the unfertilized control. What are the median heights of each of the three varieties? What fraction of the sample with fertilizer B is more than 1.5m tall?

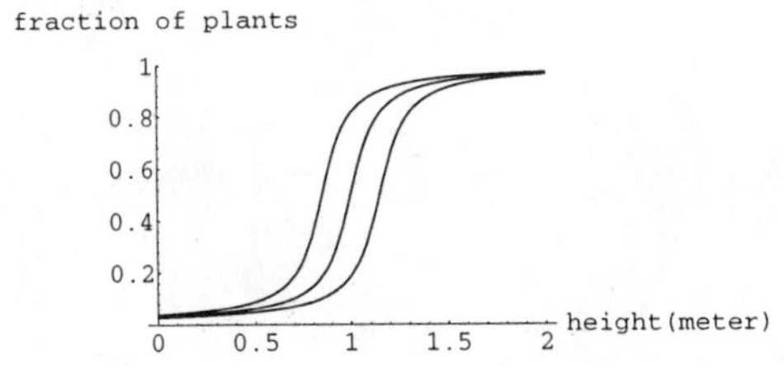

Cumulative distribution function of heights of plants

15. Sketch the polar graph $r = 5\sin\theta$ for $0 \le \theta \le \pi$.

16. Find the interval of convergence of the power series $\sum_{n=0}^{\infty} n3^n x^n$.

17. Determine if the series

$$\sum_{k=2}^{\infty}(-1)^{k+1}\frac{\ln k}{k^3}$$

converges or diverges. Explain your reasoning. If convergent, determine how large n has to be so that the partial sum S_n approximates the sum S with an error less than 0.01.

18. (a) Compute the Taylor polynomial of degree 2 for $\sqrt{1+x}$, centered at $x = 0$.

(b) The electrical potential V, at a distance r along the axis perpendicular to the center of a charged disc of radius a and charge density σ, is

$$V = 2\pi\sigma(\sqrt{r^2+a^2}-r).$$

If r is large in comparison to a, then a/r is close to 0. Use part (a) to give an approximation for

$$\sqrt{r^2+a^2}-r$$

as r times a polynomial in powers of a/r.

Final Examination for Calculus 3: Test 1

1. Consider the solid above the cone $z = 3\sqrt{x^2 + y^2}$ and below the paraboloid $z = 4 - x^2 - y^2$. Set up an integral expression to answer each of the following questions. DO NOT EVALUATE.

   (a) Find the volume of the solid.

   (b) Find the average distance from a point in the solid to the origin.

2. Suppose an area of rolling terrain can be modeled by the equation $f(x, y) = \sin(\pi x + 2\pi y)$, where $f(x, y)$ is the elevation in hundreds of feet for some point east a distance of x (hundreds of feet) and north a distance of y (hundreds of feet) from a fixed point $(0, 0)$. See the figure below. Suppose you are located 50 feet east and 75 feet north of $(0, 0)$, i.e., $x = 5$, $y = 0.75$.

   (a) What is your elevation?

   (b) If you traveled east from your location, how steep is the incline? Show this fact on the figure given.

   (c) If you traveled north from your location, how steep is the incline? Show this fact on the figure given.

   (d) Which direction is steeper and why?

   (e) Compute the tangent plane to this surface at the point $(0.5, 0.75)$.

   (f) Use the tangent plane to approximate the elevation at $(0.6, 0.8)$.

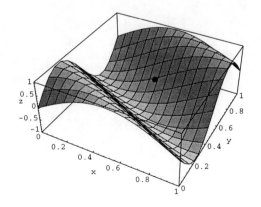

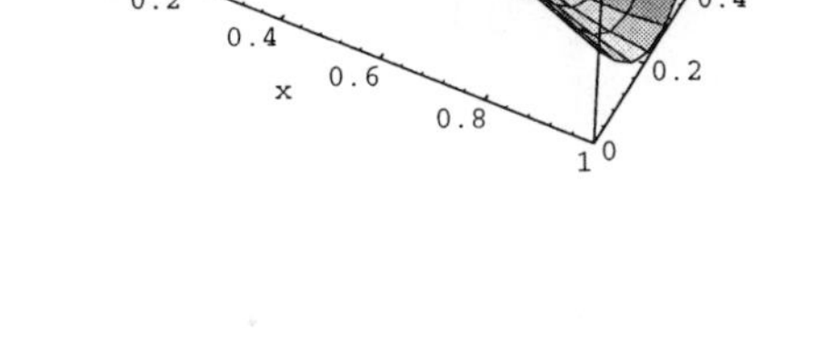

3. The depth of a lake at the point (x, y) is given by $f(x, y) = 300 - 3x^2 - 2y^2$. A boy, who is not a very good swimmer, is in the water at $(8, 2)$.

   (a) Find the rate at which the depth changes if he swims in the direction of the vector $\vec{i}$.

   (b) Find the rate at which the depth changes if he swims toward the point $(7, 4)$.

   (c) In what direction should he swim in order for the depth of the water to decrease most rapidly?

   (d) On the level curves given below, sketch in the path the boy swims, starting at $(8, 2)$, if he always swims in the direction where the depth of the water decreases most rapidly.

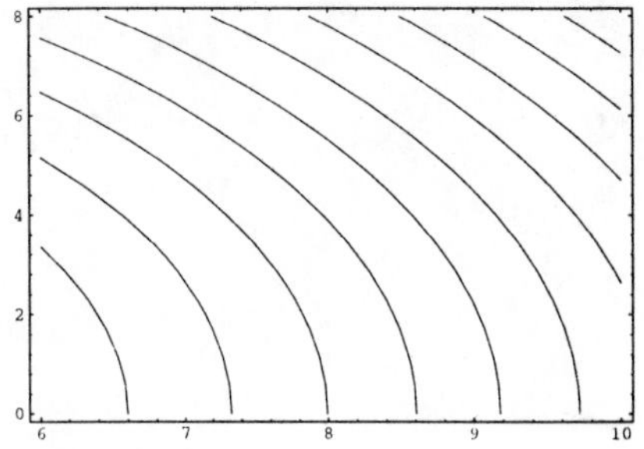

4. Find the point(s) on the surface $x^2 - yz = 1$ that is (are) closest to the point $(0, 1, 2)$.

5. Consider the function $f(x, y) = 2x^3 + 3y^2 - 6xy$.

   (a) Find all relative maximums, relative minimums, and saddle points.

   (b) Label these points on the level curves given below.

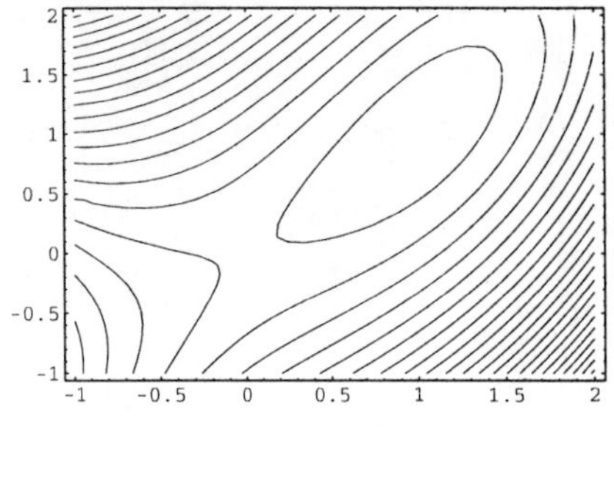

6. Let $T(x, y) = xy - x - y$ for all points in the triangle with vertices at $(0, 0)$, $(0, 1)$ and $(2, 0)$.

   (a) Find the hottest and coldest points in the triangle.

   (b) Find the average temperature in the triangle.

7. An astronaut is flying along the path described by $\mathbf{r}(t) = (t^2 - t,\ 2 + t,\ -3/t)$, where t is given in hours. At the point $(6, 5, -1)$ she shuts off the engines. Where is she 2 hours later?

8. (a) Consider the triple integral

$$\int_0^{2\pi} \int_1^2 \int_0^5 r^2 \, dz \, dr \, d\theta.$$

   Write a problem whose answer is this integral. Be sure to give a geometric description of the solid.

   (b) Find the volume of the solid in the first octant bounded by the plane $2x + y + z = 10$.

   (c) Switch the order of integration of

$$\int_0^1 \int_0^{2x} \cos(x^2) \, dy \, dx$$

   to $dx\,dy$. Which integral is easier to compute? Explain why. (You need not actually compute either integral to answer the question.)

9. (a) What do the directional derivative and the gradient of a function have to do with one another?

   (b) For which direction is the directional derivative of a function a maximum, a minimum, 0, and 1/2 its maximum value?

Final Examination for Calculus 3: Test 2

1. Evaluate
$$\iint_R e^{x^2+y^2}\, dy\, dx,$$
where R is the semicircular region bounded by the x-axis and the curve $y = \sqrt{1-x^2}$.

2. Convert $\int_0^{2\pi}\int_0^1\int_0^{\sqrt{4-r^2}} r^2\, dz\, dr\, d\theta$ to
   (a) rectangular coordinates.
   (b) spherical coordinates.

3. The volume of a solid is
$$\int_0^2\int_0^{\sqrt{2x-x^2}}\int_{-\sqrt{4-x^2-y^2}}^{\sqrt{4-x^2-y^2}} r^2\, dz\, dy\, dx$$
   (a) Describe the solid by giving equations for the surfaces that form its boundary.
   (b) Convert the integral to cylindrical coordinates but do not evaluate the integral.

4. Find the volume of the portion of the solid cylinder $x^2 + y^2 \leq 1$ that lies between the planes $z = 0$ and $x + y + z = 2$.

5. A hemispherical bowl of radius 5 cm is filled with water to within 3 cm of the top. Set up an integral for the volume of water in the bowl.

6. Let D be the region in xyz-space defined by the inequalities $1 \leq x \leq 2$ and $0 \leq xy \leq 2$, and $0 \leq z \leq 1$. Evaluate
$$\iiint_D (x^2 y + 3xyz)\, dx\, dy\, dz$$
by applying the transformation $u = x$, $v = xy$, $w = 3z$ and integrating over an appropriate region G in uvw-space.

7. Use Green's theorem to find the counterclockwise *circulation* and outward *flux* for the field
$$\mathbf{F} = (2xy + x)\,\mathbf{i} + (xy - y)\,\mathbf{j}$$
and the curve C where C is the square bounded by $x = 0$, $x = 1$, $y = 0$, $y = 1$.

8. Sketch the vector field $\mathbf{F} = (x - y)\,\mathbf{i} + (x + y)\,\mathbf{j}$ in the xy-plane.

9. Apply Green's theorem to evaluate the line integral
$$\int_C y^2\, dx + x^2\, dy,$$
where C is the circle $x^2 + y^2 = 4$.

10. Find the mass of a thin wire lying along the curve
$$\mathbf{r} = \sqrt{2}t\,\mathbf{i} + \sqrt{2}t\,\mathbf{j} + (4 - t^2)\,\mathbf{k}, \quad 0 \leq t \leq 1,$$
if the density is $\rho = 3t$.

11. The temperature at the point (x, y) on a metal plate is given by $T(x, y) = 100e - (x^2 + y)/2$, where $x \geq 0$ and $y \geq 0$. A graph of the temperature function is shown below.

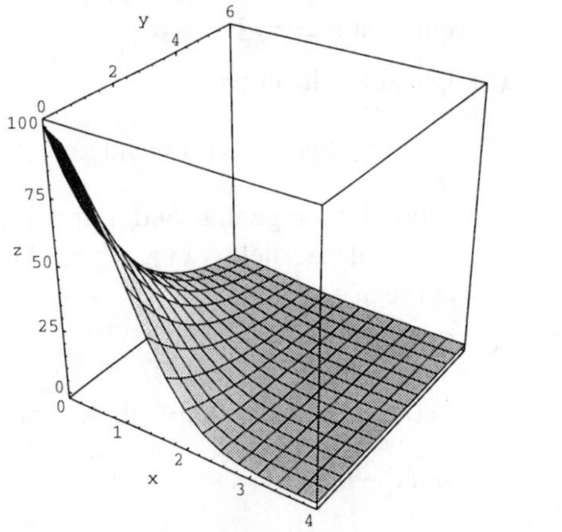

   (a) Find the direction of greatest increase in temperature from the points $(3, 5)$, $(2, 2)$, and $(1, 1)$ and draw these directions on a coordinate grid.
   (b) Find the equation of the path followed by a heat-seeking missile placed at the point $(3, 5)$ on the metal plate. Draw this path in the grid used for part (a).
   (c) Will the heat-seeking missile find the hottest spot on the plate or will it miss it? Explain what happens here.

Final Examination for Calculus 3: Test 3

1. Is the function $\{3x^2\cos(x^3+y), \cos(x^3+y)\}$ the gradient of a function? If so, what function?

2. Find the equation of the tangent and normal lines to the curve $x^2 - y^2 = 3$ at the point $(2, 1)$.

3. Consider the function

$$\frac{1}{(x-1)^2 + y^2 + 1}.$$

   Convince me of two things: First, the function has a maximum but no minimum. Second, you can find that minimum point. (The most convincing argument for the second part would be for you to show me what the minimum point is.)

4. (a) Consider the curve

$$x(t) = \cos t, \quad y(t) = \sin t, \quad z(t) = t.$$

   Find the equation of the tangent line to the curve at $t = \pi/3$.

   (b) Consider the curve

$$x(t) = \cos t, \quad y(t) = \sin t, \quad z(t) = t.$$

   Find the tangential and normal components of the field $\{xyz, x^2y^2, z^3\}$ to that curve at $t = \pi/4$.

   (c) Consider the curve

$$x(t) = \cos t, \quad y(t) = \sin t, \quad z(t) = t,$$

   and the field

$$\mathbf{f}[x, y, z] = \{x, -y, -z\}.$$

   Is the tangent vector to the curve working with or against the field at $t = \pi/4$? Is the tangent vector to the curve working with or against the field at $t = 3\pi/4$?

5. Let C denote the boundary of the elliptical region $x^2 + (y/2)^2 \leq 3$. Let $\mathbf{F}(x, y) = \{x^2, xy^2\}$. Calculate the flux, $\int_c \mathbf{F} \cdot d\mathbf{N}$, and the circulation, $\int_c \mathbf{F} \cdot d\mathbf{T}$, of this field.

6. Find the equation of the plane through the points $\{1, 2, 3\}$, $\{1, 1, 1\}$, and $\{1, 0, 2\}$.

7. Find the equation of the tangent plane to the surface $x^2 + (y/2)^2 + (3z)^2 = 11$ at the point $\{1, 2, 1\}$.

8. Show that the curves $x^2 - y^2 = A$ and $xy = B$ are orthogonal to each other at all points in the plane, except the point $(0, 0)$. (Hint: At each point in the plane, except for $(0, 0)$, you can find A and B so that these level curves pass through that point.)

9. (a) A metal gadget is to be cut out of a sheet of metal. Its boundaries are the curves $xy = 1$, $xy = 3$, $x = y/2$, $x = 3y/2$. Set up an integral for calculating the area of the gadget.

   (b) Write down the integrals needed to calculate the centroid of the gadget in part (a).

10. (a) Set up the integral in good coordinates for the volume of the region bounded by the surfaces $x + y + z = 1$, $x + y + z = 4$, $x - y + z = -1$, $x - y + z = 3$, $z - x^2 - y^2 = 0$, and $z - x^2 - y^2 = 3$. Set up, but don't try to compute, the appropriate volume conversion factor.

    (b) Find the mass of the region if the density at any point is given by

$$\rho(x, y, z) = x^2 + 2y^2 + 3z^2.$$

Final Examination for Calculus 3: Test 4

The use of calculators is not permitted on this exam. You may use the formulas for Green's theorem, the Divergence theorem, and Stokes's theorem, which are attached.

1. Find the tangent plane to the surface $y^3z - xz = 0$ at the point $P(1, 1, 1)$.

2. Find the unit tangent vector $\mathbf{T}$ to the curve $\mathbf{r}(t) = (3t^2 - 2)\,\mathbf{i} + (2t - 1)\,\mathbf{j} + e^{t-1}\,\mathbf{k}$ at $t = 1$.

3. Is the vector $\mathbf{T}$ of (b) tangent to the surface in (a) at the point $P(1, 1, 1)$? Why? (*No partial credit for yes or no without explanation.*)

4. Let $f(x, y)$ be a differentiable function. Assume that $\dfrac{\partial f}{\partial y}(1, 1) = 2$ and $\dfrac{\partial^2 f}{\partial x^2}(1, 1) = -1$. Then close to the point $(1, 1)$, the function f is:

   (a) Increasing in the direction of the y-axis and concave up in the direction of the x-axis.

   (b) Decreasing in the direction of the x-axis and concave up in the direction of the x-axis.

   (c) Increasing in the direction of the y-axis and concave down in the direction of the x-axis.

   (d) Decreasing in the direction of the x-axis and concave down in the direction of the x-axis.

5. If $z = f(x, y)$, $x = X(s, t)$, $y = Y(x, t)$, then $\partial z/\partial s$ is given by

   (a) $\dfrac{\partial z}{\partial s} = \dfrac{\partial f}{\partial s}\dfrac{\partial X}{\partial s} + \dfrac{\partial f}{\partial s}\dfrac{\partial Y}{\partial s}$.

   (b) $\dfrac{\partial z}{\partial s} = \dfrac{\partial f}{\partial x}\dfrac{\partial X}{\partial s} + \dfrac{\partial f}{\partial y}\dfrac{\partial Y}{\partial s}$.

   (c) $\dfrac{\partial z}{\partial s} = f\dfrac{\partial X}{\partial s} + f\dfrac{\partial Y}{\partial s}$.

   (d) $\dfrac{\partial z}{\partial s} = \dfrac{\partial f}{\partial s}\dfrac{\partial s}{\partial x} + \dfrac{\partial f}{\partial s}\dfrac{\partial s}{\partial y}$.

6. The extreme points of the function $f(x, y) = x^2 + y$ on the circle $x^2 + y^2 = 1$ are solutions of the system of equations

   (a) $\begin{cases} 2x &= 2x\lambda \\ 1 &= 2y\lambda \\ x^2 + y &= 0 \end{cases}$

   (b) $\begin{cases} 2x &= 2x\lambda \\ 1 &= 2y\lambda \\ x^2 + y - 1 &= 0 \end{cases}$

   (c) $\begin{cases} 2x &= 2x\lambda \\ 2y &= 2y\lambda \\ x^2 + y - 1 &= 0 \end{cases}$

   (d) $\begin{cases} 2x &= 2x\lambda \\ 2y &= 2y\lambda \\ x^2 + y &= 0 \end{cases}$

7. The level curves and the gradient of $z = f(x, y)$, $x \in [0, 2]$, $y \in [0, 2]$ are plotted in the following picture.

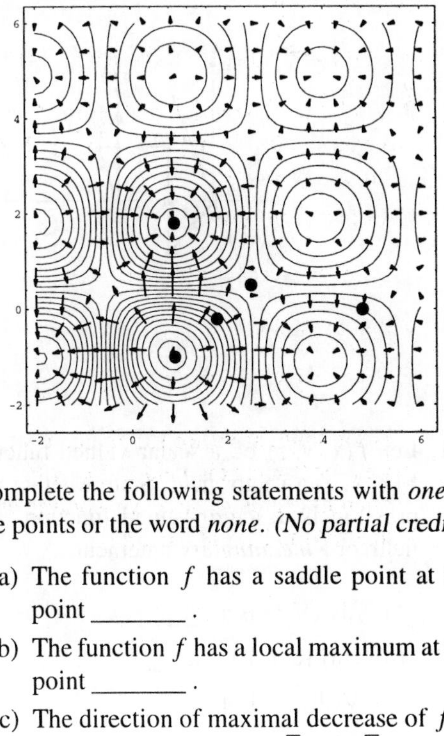

Complete the following statements with *one* of the points or the word *none*. (*No partial credit.*)

   (a) The function f has a saddle point at the point _______ .

   (b) The function f has a local maximum at the point _______ .

   (c) The direction of maximal decrease of f at the point _______ is $(\sqrt{2}/2, \sqrt{2}/2)$.

8. Let $f(x, y) = x^2 + y^2 - \frac{1}{2}x^2y$. Find all the critical points of f and classify them.

9. A vector field $\mathbf{F}$ is plotted in the figure on page 127.

   Complete the following statements with *positive, negative,* or *zero. (No partial credit.)*

   (a) The flux of f across C_1 is _______ .

   (b) The flux of f across C_2 is _______ .

   (c) The flux of f across C_3 is _______ .

   Complete the following statements with *one* of the points P_1, P_2, P_3 (or the word *none*).

   (d) The divergence of f at the point _______ is positive .

   (e) The divergence of f at the point _______ is negative.

   (f) The divergence of f at the point _______ is zero.

10. Let $\mathbf{F}(x, y, z) = y\,\mathbf{i} + x\,\mathbf{j} + \mathbf{k}$.

    (a) Find a potential for $\mathbf{F}(x, y, z)$.

    (b) Find $\int_{C_1} \mathbf{F} \cdot d\mathbf{s}$, where C_1 is given by the following figure.

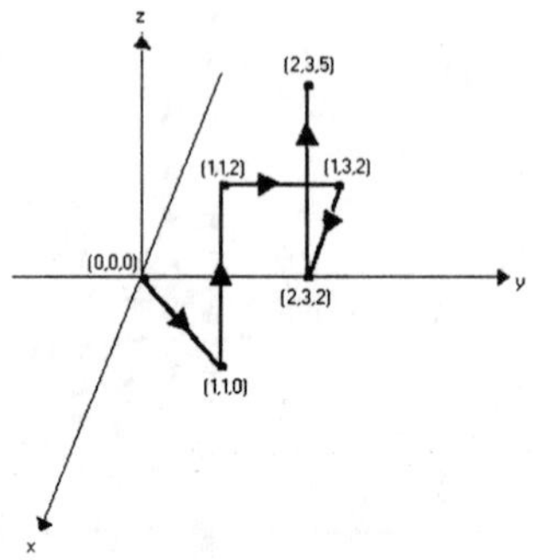

11. Let $f(x, y, z)$ be a scalar valued function and $\mathbf{F}(x, y, z)$ a vector field. State whether each expression is a *scalar*-valued function, a *vector* field, or a *meaningless* statement.

    (a) div (∇f) is a _______ .

    (b) curl (curl $\mathbf{F}$) is a _______ .

    (c) ∇(div f) is a _______ .

    (d) div (curl (∇f)) is a _______ .

12. The integral

$$\int_0^{2\pi} \int_0^{\sqrt{2}} \int_r^2 r\,dz\,dr\,d\theta$$

    represents the volume

    (a) enclosed by the paraboloid $z = x^2+y^2$ and the plane $z = 2$.

    (b) enclosed by the paraboloid $z = x^2+y^2$ the cylinder $x^2 + y^2 = 2$, and the xy-plane.

    (c) enclosed by the cone $z = \sqrt{x^2 + y^2}$ and the plane $z = 2$.

    (d) enclosed by the cone $z = \sqrt{x^2 + y^2}$ the cylinder $x^2 + y^2 = 4$, and the xy-plane.

13. The work done by the force

$$\mathbf{F}(x, y) = x(x + y)\,\mathbf{i} + xy^2\,\mathbf{j}$$

    in moving a particle along the unit circle in a counterclockwise direction is given by

    (a) $\displaystyle\iint_{x^2+y^2\le 1} \text{div}(\mathbf{F})\,dx\,dy.$

    (b) $\displaystyle\int_0^{2\pi} \mathbf{F}(\cos t, \sin t) \cdot \big[(-\cos t)\,\mathbf{i} + (\sin t)\,\mathbf{j}\big]\,dt.$

    (c) $\displaystyle\iint_{x^2+y^2\le 1} \text{div}(\mathbf{F})\,dx\,dy.$

    (d) $\displaystyle\int_0^{2\pi} \mathbf{F}(\cos t, \sin t) \cdot \big[(\cos t)\,\mathbf{i} + (\sin t)\,\mathbf{j}\big]\,dt.$

14. Let the solid D be given by the following figure.

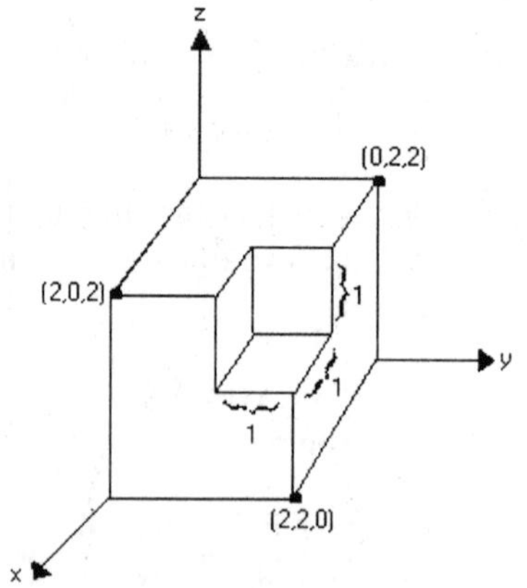

    (a) Find the volume of the solid D. Use any method. Show your work.

    (b) Let S be the surface boundary of D. Let $\mathbf{F}(x, y, z) = 2x\,\mathbf{i} + z^2\,\mathbf{j} + 3z\,\mathbf{k}$. Find the outward flux of $\mathbf{F}$ across S.

15. Find the outward flux of $\mathbf{F}(x, y, z) = x\,\mathbf{i} + y\,\mathbf{j} + 3\,\mathbf{k}$ across the piece of the paraboloid $z = x^2 + y^2$ in the first octant, bounded by the planes $z = 4$, $x = 0$, and $y = 0$.

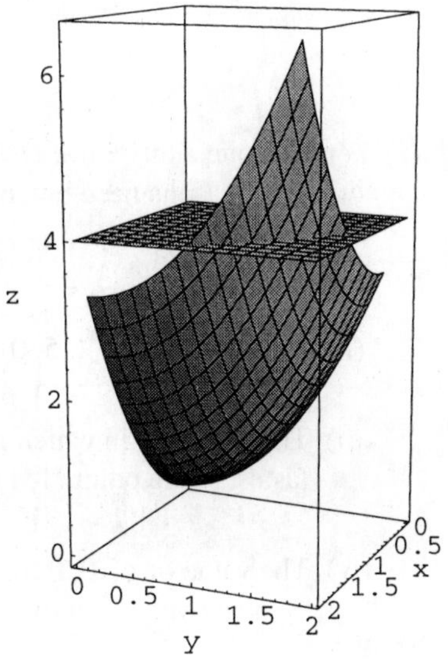

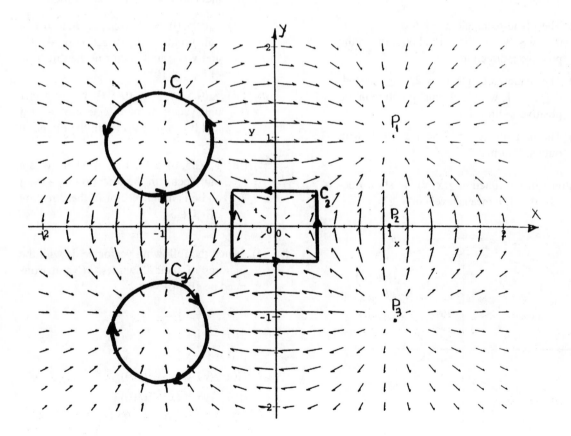

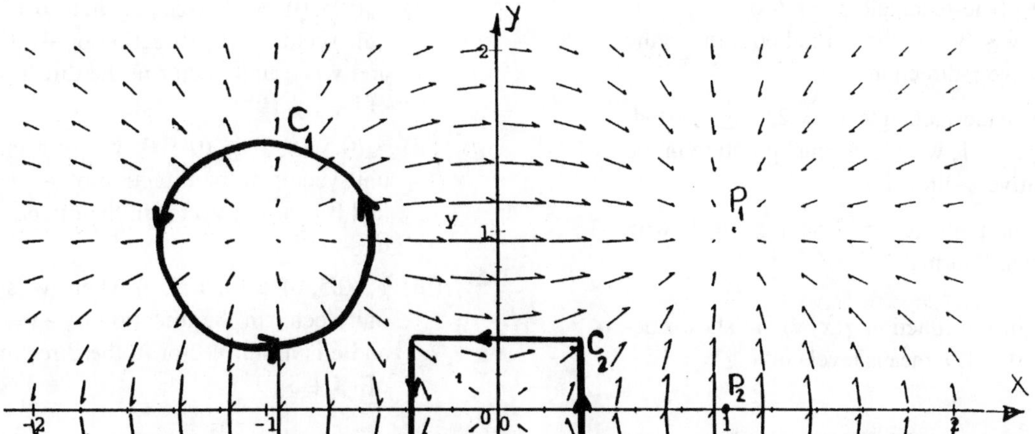

Final Examination for Calculus 3: Test 5

Calculators are allowed.

1. Calculate the line integral of the following vector fields along the given curves:

   (a) $2x\,\mathbf{i} - 3\,\mathbf{j} + 2z\,\mathbf{k}$ along the portion of the parabola $y = x^2$, $z = 1$ from $(0, 0, 1)$ to $(2, 4, 1)$.

   (b) $y\,\mathbf{i} - x\,\mathbf{j}$ counterclockwise around the circle $x^2 + y^2 = 4$.

   (c) $(x + y)\,\mathbf{i} + x\,\mathbf{j}$ along the path that goes in straight line segments from $(0, 0)$ to $(0, 4)$ to $(1, 4)$ to $(1, 1)$.

2. Calculate the flux of $\mathbf{F} = yx^2\,\mathbf{i}$ through the following surfaces.

   (a) The plane rectangle $z = 2$, $0 \le x \le 1$, $0 \le y \le 3$, with its normal pointing in the positive z-direction.

   (b) The plane rectangle $x = 2$, $0 \le y \le 4$, $0 \le z \le 1$, with its normal pointing in the positive x-direction.

   (c) The unit sphere, $x^2 + y^2 + z^2 = 1$, with outward normal.

3. Contours of the function $g(x, y)$ are shown below. [NOTE: L:c means level curve $g = c$.]

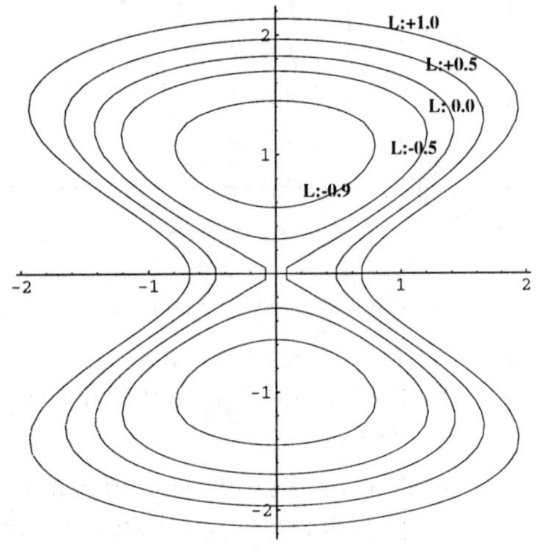

   (a) For each question below, circle the best answer. No reasons need be given.

   (i) The value of $g(1, 1)$ is approximately:
   -0.9 $\quad -0.5 \quad 0 \quad -1.0$

   (ii) The value of $g(-0.5, 0)$ is:
   negative zero positive

   (iii) The direction in which g is increasing fastest at the point $(1, 1)$ is:
   $-\mathbf{i} \quad -\mathbf{i}+\mathbf{j} \quad \mathbf{j} \quad \mathbf{i} \quad \mathbf{i}-\mathbf{j}$

   (iv) The value of $g_y(0, 0.3)$ is:
   negative zero positive

   (b) **T** (true) or **F** (false) beside each of the statements below. No reasons needed.

   (i) $g_{\mathbf{u}}(0.5, 0) > g_{\mathbf{v}}(0.5, 0)$ where $\mathbf{u}$ is a unit vector in the direction of $-\mathbf{i} + \mathbf{j}$ and $\mathbf{v}$ is a unit vector in the direction of $\mathbf{i} + \mathbf{j}$.

   (ii) $g_{\mathbf{a}}(0.5, 0) > g_{\mathbf{b}}(0.5, 0)$ where $\mathbf{a}$ is a unit vector in the direction of $-\mathbf{i} + \mathbf{j}$ and $\mathbf{b}$ is a unit vector in the direction of $-\mathbf{i} + 5\,\mathbf{j}$.

   (iii) $g_{\mathbf{w}}(0.5, 0) > g_{\mathbf{c}}(0.5, 0)$ where $\mathbf{w}$ is a unit vector in the direction of $-\mathbf{i} - \mathbf{j}$ and $\mathbf{c}$ is a unit vector in the direction of $-\mathbf{i} + 5\,\mathbf{j}$.

4. (a) Match the following vector fields with the pictures of their flowlines. (No reasons need be given.)

   (i) $y\,\mathbf{i} + x\,\mathbf{j}$

   (ii) $-y\,\mathbf{i} + x\,\mathbf{j}$

   (iii) $x\,\mathbf{i} + y\,\mathbf{j}$

   (iv) $-y\,\mathbf{i} + (x + y/10)\,\mathbf{j}$

   (v) $-y\,\mathbf{i} + (x - y/10)\,\mathbf{j}$

   (vi) $(x - y)\,\mathbf{i} + (x - y)\,\mathbf{j}$

   (b) Put arrows on the flowlines indicating the direction of flow.

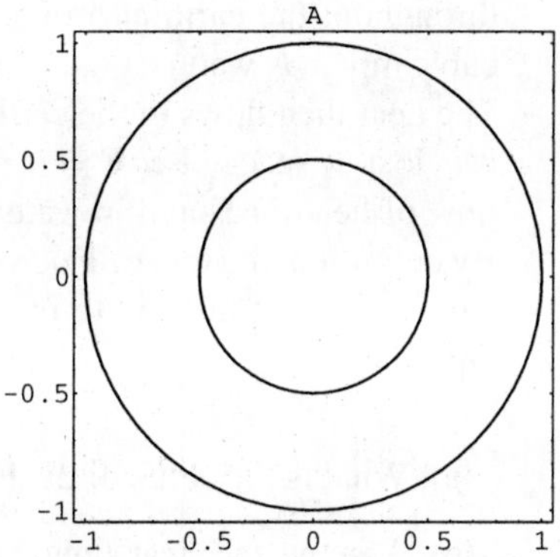

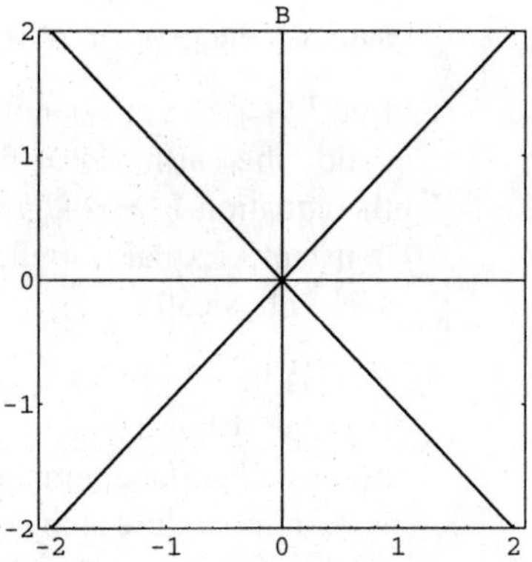

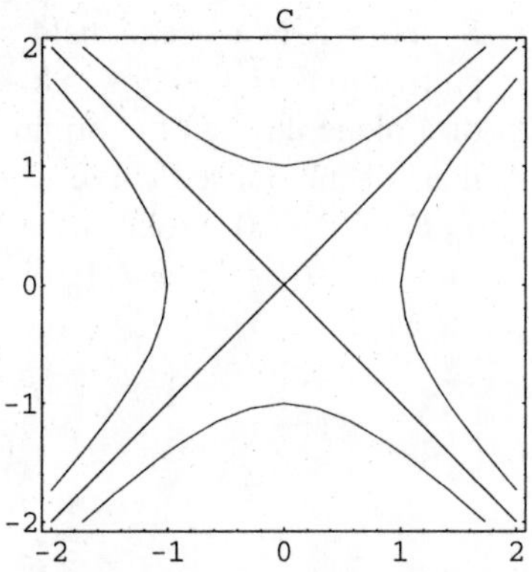

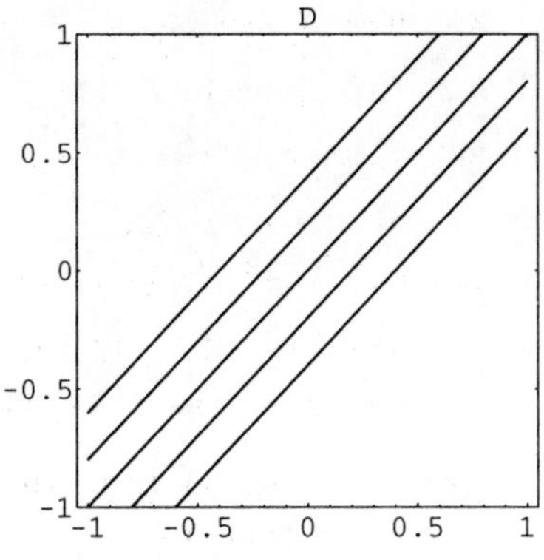

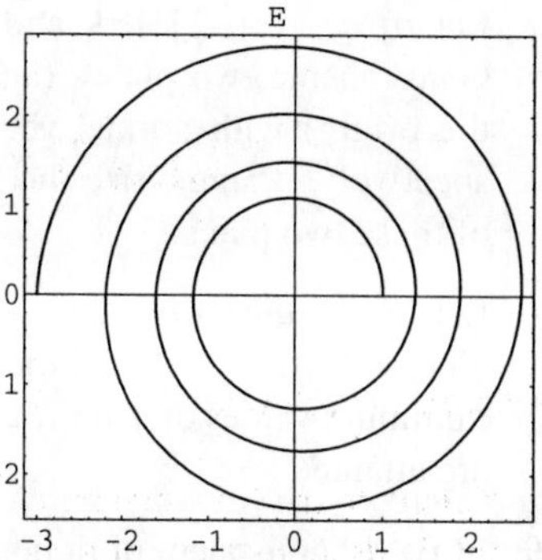

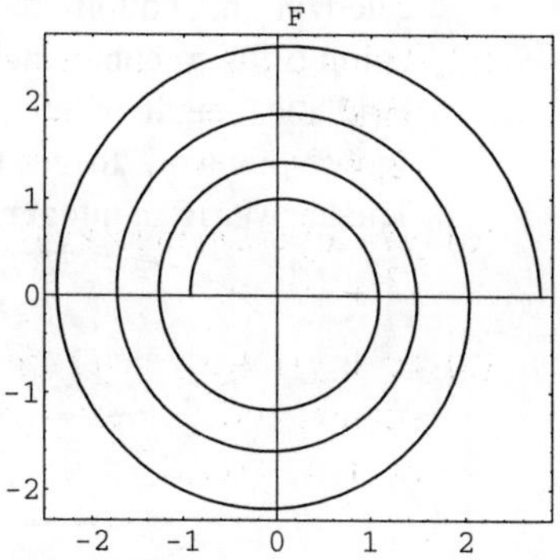

5. For each integral fill in the limits of integration so that the equation is true. V_1 is the volume of the cone $z^2 = x^2 + y^2$ that is below the plane $z = 1$ and above the plane $z = 0$.

$$V_1 = \int_0^{2\pi} \int_{\square}^{\square} \int_{\square}^{\square} \rho^2 \sin\phi \, d\rho \, d\phi \, d\theta$$

$$V_1 = \int_0^{2\pi} \int_{\square}^{\square} \int_{\square}^{\square} r \, dz \, dr \, d\theta$$

V_2 is the solid tetrahedron determined by the points $(0, 0, 1)$, $(0, 1, 0)$, $(1, 0, 0)$ and the origin.

$$V_2 = \int_{\square}^{\square} \int_{\square}^{\square} \int_{\square}^{\square} dz \, dy \, dx$$

6. Let S be the surface $z = x^2 + y^2$ above the region $R : 1 \le x \le 2, \ 1 \le y \le 2$.

   (a) Find a vector normal to S at the point $(1, 1, 2)$.

   (b) Find an equation for the tangent plane to S through $(1, 1, 2)$.

   (c) Find the area of the portion of the tangent plane found in part (b) that lies above R.

7. Let $\mathbf{n}_1 = \mathbf{i} + 2\mathbf{j} + \mathbf{k}$ and $\mathbf{n}_2 = -3\mathbf{i} - 5\mathbf{j}$. Consider the two planes (both passing through the origin) with normal vectors $\mathbf{n}_1$ and $\mathbf{n}_2$ respectively. Parametrize the line of intersection of these two planes.

8. Let R be the curve $2x^2 + y^2 = 1$ and let $f(x, y) = e^{2x^2 - xy + y^2}$. Find the maximum and minimum values of f on R and state where they are attained.

9. (a) A hemisphere of radius six inches is truncated at the bottom to make a bowl of volume 99 π cubic inches (see picture). Find the height of the bowl. You must show your work to get full credit. (Hint: The answer is an integer.)

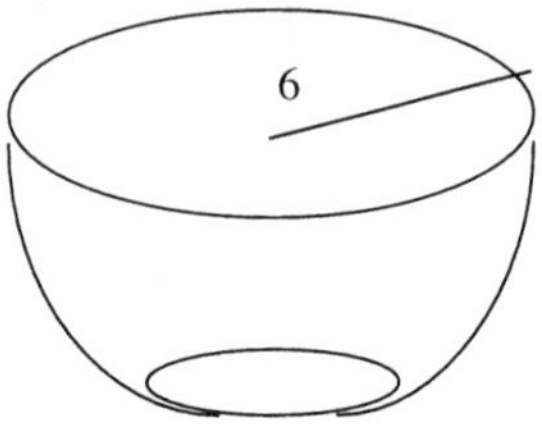

(b) Explain why your answer to part (a) is unique.

10. Let $\mathbf{F}$ be a vector field in 3-space such that on the surface of the unit sphere $\mathbf{F} = z^3\,\mathbf{k}$. If $\operatorname{div}\mathbf{F} = a$ is a constant in all of 3-space, what is the value of a?

11. Heat is generated inside the earth by radioactive decay. Assume it is generated uniformly throughout the earth at a rate of 120 watts per cubic mile. (A watt is a rate of heat production.) The heat then flows to the earth's surface where it is lost to space. Let $\mathbf{F} = (x, y, z)$ denote the flow of heat measured in watts per square mile. By definition, the flux of $\mathbf{F}$ across a surface is the amount of heat flowing through the surface per unit time.

(a) What is the value of $\operatorname{div}\mathbf{F}$? Include units.

(b) Assume the heat flows outward symmetrically. Verify that $\mathbf{F} = \alpha\mathbf{r}$, where $\mathbf{r} = x\,\mathbf{i} + y\,\mathbf{j} + z\,\mathbf{k}$ and α is a suitable constant, satisfies the given conditions. Find α.

(c) Let $T = (x, y, z)$ denote the temperature inside the earth. Heat flows according to the equation $\mathbf{F} = -k \operatorname{grad} T$, where k is a constant. Explain briefly why this makes sense physically.

(d) If T is in °C, then $k = 50,000$. Assuming the earth is a sphere with a radius 4000 miles and surface temperature 20°C, what is the temperature at the center? Show your work.

12. Suppose $\mathbf{F}$ is a vector field such that at every point $\operatorname{curl}\mathbf{F} = \mathbf{i} + 2\,\mathbf{j} + 5\,\mathbf{k}$. Find the equation of a plane through the origin with the property that for any closed curve C lying in the plane $\int_C \mathbf{F} \cdot d\mathbf{r} = 0$. Your answer should be in the form $ax + by + cz = d$. Explain your reasoning.

Final Examination for Calculus 3: Test 6

1. Below is the graph of $y = f(x)$ with $0 \le x \le 2\pi$. Using this same function f sketch a graph in polar coordinates for $r = f(\theta)$.

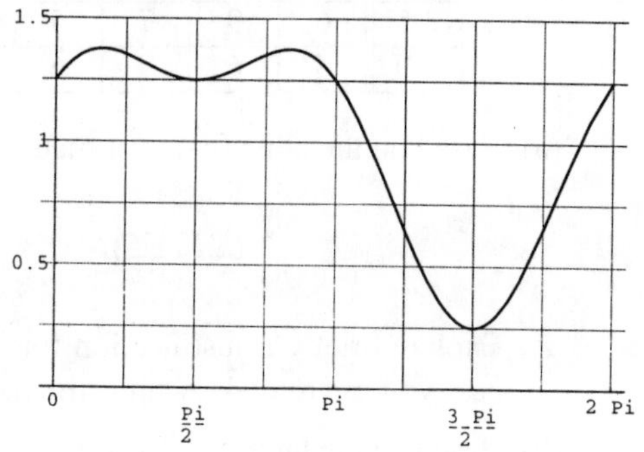

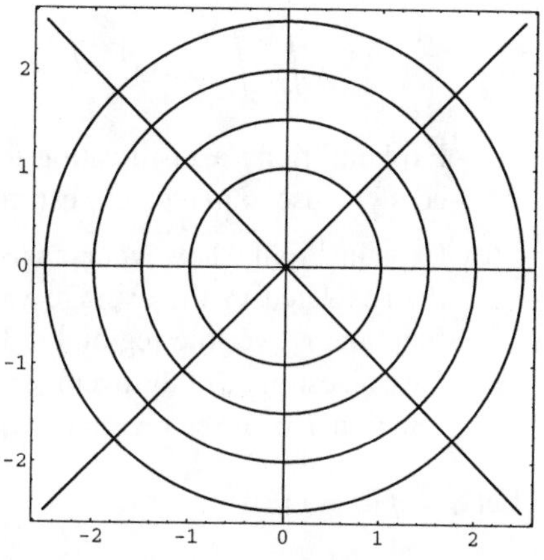

2. Suppose $\nabla f(x, y) = (y^2 + x, 2xy)$ for all (x, y) and $f(1, 0) = 5$. Find $f(2, 1)$.

3. Evaluate the following integral:

$$\int_0^2 \int_{-2}^3 \int_0^{x^2+y^2} x \, dz \, dy \, dx$$

4. Suppose $\mathbf{r}$ is a twice differentiable parametrization of a planar curve with $\mathbf{r}(t)$ being the position at time t with $0 < t < 10$ of an object moving in the plane with velocity, $\mathbf{r}'(t)$ and acceleration $\mathbf{r}''(t)$. Explain why if the speed of the object is largest at time t^* then $\mathbf{r}'(t^*) \cdot \mathbf{r}''(t^*) = 0$. (Recall that the speed is magnitude of the velocity vector.)

5. We say that $f(x, y)$ is a probability density function over the region R in the plane if

$$\iint_R f(x, y) \, dA = 1.$$

Consider $f(x, y) = kye^x$ and the rectangular region R with $0 \le x \le 2$ and $0 \le y \le 1$. Find the value of k so that f is a probability density function.

6. Use a change in the order of integration to evaluate the iterated integral

$$\int_0^1 \int_{x/2}^{1/2} e^{y^2} \, dy \, dx.$$

[HINT: Draw the figure for these limits in the plane.]

7. Set up, but *do not evaluate*, a double or triple integral to compute the volume contained between the plane $z = y + 1$, the cylinder $x^2 + y^2 = 1$, and the plane $z = 0$.

8. Find the volume of the solid contained between the paraboloid $z = 8 - x^2 - y^2$ and the plane $z = 4$ using polar or cylindrical coordinates. [HINT: The sketch of this solid should help to find the appropriate limits of integration.]

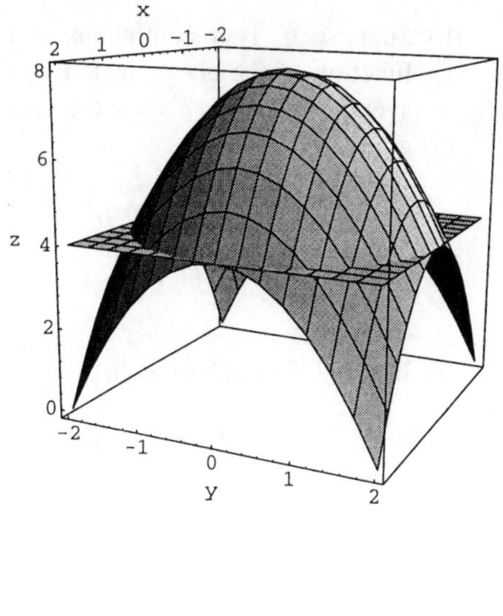

9. An object is moving in a coordinate plane so that its velocity at time t seconds is

$$(5t^2 + 3t - 4, 1 - t) \text{ meters/second.}$$

At time 2 the object is at the point with coordinates $(1, 0)$.

(a) Find the position of the object at time $t = 0$.

(b) Express the distance traveled by the object during the time interval $[0, 2]$ as an integral but DO NOT EVALUATE.

10. Find parametric equations for the line that passes through the point $(2, 3, 1)$ and is parallel to the line of intersection of the planes with equations $2x - 2y - 3z = 2$ and $3x - 2y + z = 1$.

[HINT: The line of intersection is perpendicular to normal lines for either plane.]

11. True or False.

(a) If $\mathbf{v} \cdot \mathbf{w} = 0$, then $\mathbf{v}$ and $\mathbf{w}$ are orthogonal vectors.

(b) $\mathbf{v} \times \mathbf{w} + \mathbf{w} \times \mathbf{v} = 0$.

(c) $\mathbf{i} \times (\mathbf{j} \times \mathbf{j}) = (\mathbf{i} \times \mathbf{j}) \times \mathbf{j}$.

(d) $|\mathbf{v} \times \mathbf{w}| = \|\mathbf{v}\| \, \|\mathbf{w}\| \sin t$ where t is the angle between $\mathbf{v}$ and $\mathbf{w}$.

(e) If $f(x, y)$ has both partial derivatives at all points in the plane, then f is a continuous function.

(f) If $f(x, y)$ is differentiable at (a, b), then it is continuous at (a, b).

(g) If θ is the angle between unit vectors $\mathbf{u}$ and $\mathbf{u}'$, then $\cos\theta = \mathbf{u} \cdot \mathbf{u}'$.

(h) Suppose f is a continuous real valued function of the planar region R, then the average value of f over the region R is given by

$$\frac{\iint\limits_R f(x, y)\, dA}{\iint\limits_R dA}.$$

(i) If $\nabla f(2, 3) = (2, 4)$, then

$$df(2, 3) = 2\, dx + 3\, dy.$$

12. Suppose that $f(x, y) = x^2 + y$.

(a) Draw the level curve for $f(x, y) = 4$.

(b) Describe the level curve $f(x, y) = k$ for a constant k.

13. Suppose the values of $z = f(x, y)$ are given by the following table.

x/y	1.5	2.0	2.5	3.0
1.5	3	5	5	7
2.0	4	8	6	4
2.5	6	9	7	5
3.0	4	6	9	6

(a) Based on the table for f estimate

$$\frac{\partial f}{\partial y}(2.5, 2.5).$$

Explain briefly a justification for the process you use to make your estimate.

(b) Based on the table for f estimate

$$\int_2^3 \int_1^3 f(x, y)\, dx\, dy.$$

Explain briefly a justification for the process you use to make your estimate.

(c) Explain briefly how your answer in part (b) is related to the average value of the function f over the region $[1, 3] \times [2, 3]$. Give an estimate of the average value of the function f over the region $[1, 3] \times [2, 3]$.

14. Let $z = f(x, y) = x^2 - 5xy$.

(a) Suppose $x = 2\cos t$, and $y = \sin t$. Use the chain rule to find dz/dt when $t = 0$.

(b) Find $\nabla f(1, 2)$.

(c) Suppose $\mathbf{u} = \frac{\sqrt{5}}{5}\mathbf{i} + \frac{2\sqrt{5}}{5}\mathbf{j}$. Find $D_{\mathbf{u}} f(1, 2)$, the directional derivative of f in the direction of $\mathbf{u}$.

(d) Find the equation of the plane tangent to the graph of $z = f(x, y)$ at the point $(1, 2, -9)$.

(e) Show that the only critical point for f is at $(0, 0)$. Is this a relative extreme for z? If so, is it a maximum or a minimum? If possible explain your conclusions using the discriminant.

Reform and AP Calculus

Ray Cannon and *Anita Solow*
BAYLOR UNIVERSITY and GRINNELL COLLEGE

The Advanced Placement Calculus program is undergoing changes at this time as part of the calculus reform movement. The first step was a technology requirement; the second, more profound change, is a redefinition of the AP calculus course. Beginning with the May 1995 examinations (both AB and BC), some questions require the use of a graphing calculator. Graphing calculators are not allowed on the first part of the multiple choice section of the exam (25 questions; 50 minutes), but are allowed on the second part of the multiple choice exam (15 questions; 40 minutes), and on the free response section (6 questions; 90 minutes).

The College Board committee that made the decision to require graphing calculators is the same committee that constructs the examinations. This committee is aware of the wide differences in capabilities among the graphing calculators, and continues to wrestle with the problem of equity that arises from these differences. In order to achieve an equitable level of technology, the committee develops examinations assuming that the only requirements on technology are that the calculator have the capability to:

- produce the graph of a function within an arbitrary viewing winder;
- find the zeros of a function;
- compute the derivative of a function numerically; and
- compute definite integrals numerically.

But even this does not level the playing field entirely; some calculators have these capabilities built in, and others do not. To overcome these differences, the committee decided to allow students to write supplementary programs into their calculators that will then give their calculators any missing capabilities. These programs are written into the calculators prior to the exams, and the calculator memories are not cleared either before or after the exam. There are some restrictions on the type of calculator permitted; computers, electronic writing pads, and QWERTY keyboards are not allowed.

For more detailed information about the exam, including a list of approved calculators and sample supplementary programs, write to, The Advanced Placement Program, PO Box 6670, Princeton, NJ 08541–6670 and request the *AP Course Description in Mathematics*.

The Changing Course Content

The introduction of the graphing calculator does not in itself "reform calculus," but it quickly leads those involved to rethink the goals of the course. The College Board has recognized this fact and is in the process of reshaping the course description. The new course description will be available in Spring 1996 from the College Board. The process of revising the course description has included dialogue with leaders in the reform movement, high school teachers, representatives for professional organizations, and others. The AP Calculus Program is very large and very diverse; last year over 123,000 students representing about 8000 high schools took one of the two examinations. Changes in such a large program require time; the new course description will apply to courses starting in Fall 1997 and govern the May 1998 exams.

The goals for the new AP Calculus course are:

- Students should be able to work with functions represented in a variety of ways: graphical, numerical, analytic, or verbal. Students should understand connections among these representations.

- Students should understand the meaning of the derivative in terms of a rate of change and local linear approximation; they should be able to use derivatives to solve a variety of problems.

- Students should understand the meaning of the definite integral both as a limit of Riemann sums and as the net accumulation of a rate of change; they should be able to use integrals to solve a variety of problems.

- Students should understand the relationship between the derivative and the definite integral as expressed in both parts of the Fundamental Theorem of Calculus.

- Students should be able to communicate mathematics both orally and in well-written sentences. Students should be able to explain solutions to problems.

- Students should be able to model a written description of a physical situation with a function, a differential equation, or an integral.

- Students should be able to use technology to help solve problems, experiment, interpret results, and verify conclusions.

- Students should be able to determine the reasonableness of solutions, including sign, size, relative accuracy, and units of measurement.

- Students should develop an appreciation of calculus as a coherent body of knowledge and as a human accomplishment.

As presented in the new course description, Calculus AB and Calculus BC are primarily concerned with developing an understanding of the concepts of calculus, and providing experience with its methods and applications. The examinations will be designed with these goals in mind. Therefore, the courses should emphasize a multi-representational approach to calculus, with concepts, results, and problems being expressed in multiple ways: geometrically, numerically, analytically, and verbally.

Broad concepts and widely applicable methods should be emphasized. The focus of the examinations is neither manipulation nor memorization of an extensive taxonomy of functions, curves, theorems, or problem types. Thus, although facility with manipulation and computational competence are important outcomes, they are not the core of these courses.

Both courses are intended to be challenging and demanding calculus courses. They differ in content, but students in both courses study many of the same topics. Calculus BC is an extension of Calculus AB, rather than an enhancement; common topics will be tested at a similar conceptual level.

Since technology is required on the examinations, it should be used regularly by students and teachers to reinforce the relationships among the multiple representations of functions, to confirm written work, to implement experimentation, and to assist in interpreting results.

Through the use of the unifying themes of derivatives, integrals, limits, approximation, and applications and modeling, the course becomes a cohesive whole, rather than a collection of unrelated topics.

Reform and the Present

Already some school teachers are using some principles of the reform movement as they prepare their students for the AP Examination. But what about texts? We would like to stress that the AP course description explains the topics that might be covered on the exam. There is no "AP Syllabus" that says when and how these topics should be covered. The object is to provide a course at a certain high level. We know of high school teachers who use "reform" materials in their classroom, and whose students do very well on the exam.

It may be that in order to cover a specific topic which could be on the exam (l'Hôpital's Rule, for example), the teacher would have to supplement a certain text, but it is easier to supplement a "lean" text than it is to help students see the important ideas in a cluttered text. Having a text that excites the teacher and the student is far more important than coverage of one or two topics. A student who has had a course that deals with the major ideas of calculus, who has been stimulated to think about these ideas, will not only enjoy the course, but will also do well on the AP Examinations.

Part IV:

Connections

Calculus does not exist in a vacuum. By its nature as one of the greatest and most powerful intellectual developments, calculus is inextricably connected to a variety of other disciplines and other portions of the mathematics curriculum.

It is a primary mathematical tool for physics, engineering, and chemistry; it is almost as valuable a tool for modern economics and biology, as well as many other fields. In high schools and two-year colleges, calculus is often regarded as the culmination of all previous mathematical experiences. In universities, it is the gateway through which students must pass to more advanced courses in mathematics and in the client disciplines.

If we are to talk intelligently about changing calculus, then we must also consider all of the other areas to which calculus is connected. If significant changes are to occur in calculus involving different skills, different expectations, different experiences, and different learning environments, then we should want our students to come into calculus from a comparably different preparatory experience. Such courses are currently being developed and implemented. We refer the interested reader to *Preparing for the New Calculus*, edited by Anita Solow (MAA Notes, Number 36) for descriptions of several such efforts and discussions of the objectives and philosophies for such new preparatory courses.

Of course, many students come to college ready for calculus, having taken all their preparatory mathematics in high school. But, the high school mathematics curriculum is changing rapidly, in large measure due to the NCTM *Standards*. In his article for this section, John Dossey describes the status of the reform movement in the secondary schools and discusses how it connects to college calculus, both traditional and reform. In addition, the AP calculus exam is currently being revised to reflect many of the ideas common to the new calculus courses at the college level. This is discussed in Part III.

Furthermore, once the calculus experience changes, it becomes essential that the courses that follow calculus be re-thought as well. Many of the new calculus projects have incorporated significant introductions to differential equations, which has major implications for the first course in differential equations. A separate reform movement has been working to develop modern approaches to these courses, as described by Bob Borrelli and Courtney Coleman in their article in this section. Similarly, there are major implications for the upper-division mathematics offerings. David Carlson and Wayne Roberts discuss that issue in their article.

Another issue connected to calculus reform is the new GRE mathematics examination that will be given beginning in 1997 to all students going on to graduate programs in any of the quantitative disciplines including mathematics, the physical and biological sciences, engineering, computer science, economics, and business. An article discussing this test, along with some sample questions that reflect the spirit of the new calculus, is included in Part III.

Finally, no discussion of the connections to calculus is complete without considering how the client disciplines react to the changes, because other departments depend so heavily on calculus. Simultaneously, major changes are happening in most of the other curricula. What implications do they have for us in mathematics? Both of these issues are addressed in an article presenting a round-table discussion among some of the leading educators from most of the disciplines that use calculus.

Weaving a Program of Secondary School Mathematics: Precursing or Reacting to Collegiate Reform?

John A. Dossey
ILLINOIS STATE UNIVERSITY

The teaching and learning of mathematics has received a great deal of attention over the past decade. At the heart of this discussion is the National Council of Teachers of Mathematics move to develop *standards* for the K–12 mathematics curriculum, its delivery, and its assessment [2, 3, 14]. The intertwined actions of the Mathematical Association of America's focus on improving the undergraduate experience in calculus [6, 15, 16, 18], and the American Statistical Association's efforts to revise the content and teaching of introductory statistics [8, 9] have signaled that major, and fundamental, changes for education in the mathematical sciences are underway at all levels.

At the collegiate level, the focus on revolution in the teaching of calculus has been much more of an evolution [17]. The focus has changed gradually over the decade, from the initial discussion of developing a "leaner and livelier" course to the integration of new pedagogical methods to the emergence of textbooks which are beginning to reflect differences in the actual arrangement of curricular content and presentation of central concepts, principles, and algorithms in new contexts. These texts are better suited to new pedagogical approaches and the use of technology in the teaching of mathematics than their predecessors were. The transformation of the teaching of calculus is evolving through changes in the content selected, the methods of delivering that content to students in the classroom, and through the development of new ways and the modification of old ways of assessing how and the degree to which students are learning the material.

Reform at the Secondary Level

The release of the NCTM *Curriculum and Evaluation Standards for School Mathematics* [3] in 1989 capped a six-year period of introspection within the mathematics community involved with K–12 education. This self-examination dealt with fundamental questions of what were the basic values for a program of study for American youth heading into the 21st century. The result was a comprehensive statement of goals which provide a framework for school mathematics programs. These goals are:

1. Learning to value mathematics.
2. Becoming confident in one's own ability.
3. Becoming a mathematical problem solver.
4. Learning to communicate mathematically.
5. Learning to reason mathematically.

These goals were then explicated through a number of statements of what students should know and be able to do. These statements, called *standards*, describe the processes and mathematics that should be used to judge the quality of a school mathematics program. Implicit in the statements about school mathematics programs is the vision of students involved in "doing mathematics," not just receiving lectures about mathematics in their classes. The listing of content clearly provides for a richer and broader program in mathematics for *all students*. The term *all students* has almost become an anthem for the *Standards*. By the term, the NCTM means that each student should have an opportunity to achieve their full potential in school mathematics, not be shunted aside at some point based on the ability to do procedural skills, and left to rot mathematically in a general mathematics class. Rather, all students should have the opportunity to continue through a core program of studies encompassing major concepts of algebra and geometry as part of their secondary school curriculum. Such a vision makes algebra a "pump and not a filter" in the secondary pipeline to the study of more advanced courses. Significant efforts are now being made to see that *all students* do get a chance to learn and apply algebra as part of their high school studies.

Curricular Reform

The focus provided by the standards also called on programs to give greater emphasis to central processes in mathematics. Similar to the examination of the procedural knowledge—conceptual understanding balance examined in the calculus reform, secondary schools were asked to re-examine their curricula and the methods used to deliver them to students. The first four standards in the listing of the mathematics program destined for students in grades 9–12 are problem solving, reasoning, communication, and making connections within mathematics and between mathematics and other disciplines. These four standards, often called the *process standards*, have shifted interest to the pedagogy employed in the classroom. "Problem solving" has translated into added emphasis on using problems as a main focal point in designing instruction and in presenting new material. "Reasoning" has translated into a re-examination of the role of proof, not its elimination, and increased emphasis on asking students to explain their patterns of thought as they seek to build convincing arguments about a given generalization as they work to justify it. "Communication" has translated into asking students to discuss mathematics, read mathematics, write about mathematics, and model mathematics as part of their learning processes. "Connecting mathematics" has translated into tying what is being studied in mathematics to past and future objects of study within the mathematics curriculum, as well as to topics in other areas of study, e.g., population settlement patterns in a history course, the speed of a reaction in a chemistry course, the collection and analysis of data reflecting an issue of current interest in a civics course, the rate of addition and loss of words in a living language in an English course, or the accumulation of value in an account being studied in a business class.

The focus in learning about algebra while studying functions has grown from an interpretation of what students should know and be able to do with their knowledge of algebra, functions, statistics, and other areas of content. It appears to follow the sequence: data, pattern, variable, function, equation, model. This sequence sees students examining data related to the content to be developed. They are encouraged to look for patterns in the data, i.e., what can be quantified, what is changing. From this attempt at quantification, a variable is identified, as are constants. This leads to the development of functional relationships. These relationships can be manipulated symbolically, graphically, and numerically to test for what values the relationship equal a particular constant. This necessitates an understanding of equation and its solution(s). The entire sequence spells out the act of modeling in mathematics. Under what assumptions does this value obtain? What happens if we modify a parameter in the functional model? How does this relationship compare and contrast with the one that was developed earlier in another setting? Clearly, this approach to developing content, found in a number of the newer materials developed since the release of the *Standards* [1, 7, 13], matches up with the approach to knowing and solving problems found in many of the calculus reform projects.

Pedagogical Reform

The *Teaching Standards* [14] were released in 1989. They made explicit what was implicit in the *Curriculum Standards*. The teaching of mathematics should follow a constructive model for learning. Students should come to know as a result of having wrestled with material in a rich conceptual setting, fitting the new concepts and generalizations into their matrix of understandings, constructing links between the new material and that previously mastered. Instruction should be centered about tasks dealing with *worthwhile mathematics*. The implication of this was that the secondary curriculum would need to become more focused and leaner as well. Asking students to learn in groups, working together to understand, with the teacher as facilitator, immediately indicates that learning will take a longer time. The central role of tasks indicated that a great deal of work needed to be done in order to find problems that were rich enough to generate deep understanding of central concepts, while still moving students forward toward an understanding of a broad range of mathematical concepts and skills. Hence secondary mathematics was facing many of the same problems inherent in the movement to reform calculus. One difference was that secondary teachers had already had faculty development in the use of graphical calculators, writing to learn, cooperative learning, and alternative assessment methods. What stood as a major challenge for them was how to weld these together as a new mode of teaching centered around tasks. Due to the lack of faculty development programs aimed at mathematics instruction at the collegiate level, collegiate teachers needed more time to develop the individual methods, as faculties, before the curricular questions could be addressed.

Assessment Reform

The final portion of the *Standards* were published in 1995—the *Assessment Standards* [2]. These recommendations call for the application of a wide variety of methods to be employed in gathering information to inform instruction and improve the learning of students. These methods include using projects and student journals, observing group work, and listening to students' conceptions of mathematical situations. Again the recommendations are similar to those that have grown out of the calculus reform movement at the collegiate level.

Evidence of Change

The degree to which the content recommendations of the *Curriculum Standards* have been adopted in secondary schools has not been carefully assessed at this point. However, a recent NSF funded status study [19] indicates that 56% of the nation's high school teachers are "well aware" of the NCTM *Curriculum Standards*, while only 39% are "well aware" of the *Teaching Standards*. As the *Assessment Standards* appeared in early summer 1995, data on teachers awareness of them is not available.

The 1992 National Assessment of Educational Progress (NAEP) [4, 5, 10, 11, 12] attempted to measure the effect of reform in a number of ways at the student, teacher, and school levels. Results from the 1992 NAEP assessment indicates that:

- Students' reports of their attitudes toward mathematics indicate additional room for improvement. Only 51% of twelfth graders agree or strongly agree with the statement "I like mathematics." While the agree/strongly agree percentage increases to 71% for "Mathematics is useful for solving everyday problems," it remains that 29% of our youth in secondary school still doubt the efficacy of our discipline. Essentially the same results, 74%, hold for agree/strongly agree with "Almost all people use mathematics in their jobs." These results indicate that a great deal of work remains to help all students see mathematics as a vital force in their lives.

- Students are reporting having taken more advanced course work in mathematics. Over 30% of the nation's eighth graders are currently enrolled in Algebra I, and nearly 50% of the students still in school at the twelfth grade level have completed Algebra II. However, less than 15% have continued on to the study of Precalculus or a more advanced course at the secondary school level. While this marks a substantial increase over the past decade, the results pale in comparison with those found in economically competitive nations.

- Teachers at the pre-secondary level are reporting movement toward a more broadly-based curriculum, one that provides student introduction to concepts in algebra, geometry, and probability and statistics.

- Students and teachers both report increased usage of calculators and computers in the learning of mathematics. At grade twelve, 92% of the students report having a calculator for their use in mathematics class. Eighty-two percent of 12th graders taking mathematics report using the calculator in mathematics class on a weekly basis. Eleven percent report never or hardly ever using the calculator in mathematics.

- Both teachers' and students' reports reflected varying emphases on newer pedagogical and assessment methods. Small-group learning methods were reported being used at least weekly in mathematics by 42% of twelfth grade students enrolled in mathematics. However, only 3% of these students indicated that they were assigned projects in their mathematics classes. The frequency of testing indicates the degree to which the focus is on long-term conceptual results or driven by short-term skill and procedural goals. Sixty-one percent of the twelfth graders report at least weekly testing. This is a 10% drop from similar data collected in 1990 and reflects movement in the correct direction. Fifteen percent of twelfth graders taking mathematics report that they are asked at least once weekly to write about how they solved a mathematics problem.

- Assessments that ask students to work on extended problems as part of their learning are rare, but progress is seen on this front. The 1992 NAEP Assessment in Mathematics employed, for the first time, problems requiring extended student-constructed responses. While performance on these problems was low [5], pilot testing results of similar problems for the 1996 NAEP indicates a potential improvement in student access to and performance on similar problems. Newer curricula available to students at the secondary level also feature such problems [1, 7, 13].

These results reflect the fact that there is as much to be done in improving secondary mathematics as collegiate mathematics [17]. However, the trend lines are encouraging. The paths between reform in secondary school mathematics and undergraduate mathematics appear similar. Students' completion of reformed programs at the secondary level are probably nearly equal to the level of such programs available at the collegiate level. The only problem is the articulation of students from the secondary level into like programs at the collegiate level. The removal of discontinuities in this transition will do a great deal to help programs move forward smoothly on each level.

Remaining Challenges

The articulation of goals and programs across the gap between secondary schools and collegiate programs remains a barrier to reform. The movement in the teaching of calculus and the welcoming of technology into lower-division courses has done a great deal to open the log jam backing up reform at the secondary level. It has also opened the door to new possibilities in the latter two years of the undergraduate experience. Most importantly, it has created a common platform for discussion about the content of mathematics courses, and the methods of teaching this content for high school and college teachers of mathematics. This sharing of views and methods will eventuate in improved articulation and shared materials and methods.

The conversation that is abroad at the moment on the desired levels of student proficiency in conceptual understanding, procedural knowledge, and problem solving ability will continue. It is clear that high schools are not rejecting all of the traditional emphases on students' ability to manipulate expressions symbolically, to prove theorems in algebra and geometry, or to study trigonometric relationships. However, the methods by which students are dealing with this material is changing. The emphasis is no longer solely on symbolic manipulation. Students are being asked to view material through a number of representations: tabular, graphical, symbolic, and verbal. This change in approach is featured in most of the reformed programs in the calculus. The emphasis is on helping students become more flexible in their problem solving and in their ability to link important mathematical concepts.

Perhaps the biggest challenge facing school and collegiate mathematics today is the definition of what should the liberally-educated citizen and the competently-educated mathematics major know at the end of the K–16 trek through the school and its attendant emphasis on mathematics? What should a student know about mathematical reasoning, about scientific computation, about the nature and history of the discipline, about modeling real-life problems with mathematical concept and principles, about communicating about mathematical situations through speaking, reading, writing, and modeling? These questions, focusing on student growth in our discipline, form the woof weaving with the warp of content strands to form the fabric of the mathematics curriculum. The acceptance and integration of common forms of reformed curricula at the pre-secondary, secondary and post-secondary levels may make the possibility of a common consideration of these questions possible for both K–12 and collegiate faculty interested in the teaching and learning of mathematics. Thus, school mathematics may come to be the precursor of collegiate mathematics in a fashion never before possible in the American experience. As school and collegiate programs move to plan together, they can ensure curricula re-designed with appropriate opportunities and sequences to bring students to an educated understanding of mathematics. It remains for the schools and the citizens of our country to see that this happens.

References

[1] *Addison-Wesley Secondary Mathematics.* Menlo Park, CA: Addison-Wesley Publishing Company, 1995.

[2] *Assessment Standards for School Mathematics.* Reston, VA: National Council of Teachers of Mathematics, 1995.

[3] *Curriculum and Evaluation Standards for School Mathematics*. Reston, VA: National Council of Teachers of Mathematics, 1989.

[4] Dossey, J.A.; Mullis, I.V. S.; Gorman, S.; Latham, A. *How School Mathematics Functions: Perspectives from the NAEP 1990 and 1992 Assessments*. Washington, DC: National Center for Education Statistics, 1994.

[5] Dossey, J.A.; Mullis, I.V.S.; Jones, C.O. *Can Students Do Mathematical Problem Solving?* Washington, DC: National Center for Education Statistics, 1993.

[6] Douglas, R.G., Editor. *Toward a Lean and Lively Calculus*. MAA Notes No. 6. Washington, DC: Mathematical Association of America, 1986.

[7] Gordon, S.P.; Gordon, F.S.; Fusaro, B.A.; Siegel, M.J.; Tucker, A.C. *Functioning in the Real World: A Precalculus Experience*. Preliminary Edition. Reading, MA: Addison-Wesley Publishing Company, 1995.

[8] Gordon, F., and Gordon, S., Editors. *Statistics for the Twenty-first Century*, MAA Notes No. 26. Washington, DC: Mathematical Association of America, 1993.

[9] Hoaglin, D.C., and Moore, D.S., Editors. *Perspectives on Contemporary Statistics*. MAA Notes No. 21. Washington, DC: Mathematical Association of America, 1992.

[10] Mullis, I.V.S.; Dossey, J.A.; Campbell, J.R.; Gentile, C.A.; O'Sullivan, C.; Latham, A.S. *NAEP 1992 Trends in Academic Progress*. Washington, DC: National Center for Education Statistics, 1994.

[11] Mullis, I.V.S.; Dossey, J.A.; Owen, E.; Phillips, G. *NAEP 1992 Mathematics Report Card for the National and the States*. Washington, DC: National Center for Education Statistics, 1993.

[12] Mullis, I.V.S.; Jenkins, F.; Johnson, E.G. *Effective Schools in Mathematics: Perspectives from the NAEP 1992 Assessment*. Washington, DC: National Center for Education Statistics, 1994.

[13] *Pacesetter Mathematics*. New York, NY: The College Board, (in press).

[14] *Professional Standards for Teaching Mathematics*. Reston, VA: National Council of Teachers of Mathematics, 1991.

[15] Steen, L.A., Editor. *Calculus for a New Century*. MAA Notes No. 8. Washington, DC: Mathematical Association of America, 1987.

[16] Solow, A.E., Editor. *Preparing for a New Calculus*. MAA Notes No. 36. Washington, DC: Mathematical Association of America, 1994.

[17] Tucker, A.C., and Leitzel, J.R.C., Editors. *Assessing Calculus Reform Efforts*. MAA Report. Washington, DC: Mathematical Association of America, 1995.

[18] Tucker, T.W., Editor. *Priming the Calculus Pump: Innovations and Resources*. MAA Notes No. 17. Washington, DC: Mathematical Association of America, 1990.

[19] Weiss, I.R.. *A Profile of Science and Mathematics Education in the United States: 1993*. Chapel Hill, NC: Horizon Research, Inc., 1994.

New Directions for the Introductory Differential Equations Course

Robert L. Borrelli and *Courtney S. Coleman*
HARVEY MUDD COLLEGE

The introductory ODE course is currently undergoing a renaissance. A great deal of thought has gone into the goals of that course and into the means for delivering those goals. The roots of all this run deep, but lately are beginning to break to the surface. As a course traditionally required in all science and engineering curricula, there is the expectation that the material covered in the course will be useful in the "client disciplines." For a long time this meant that the course was centered mostly on the theory and techniques for finding solution formulas, a quick introduction to numerical solution methods, and a few traditional applications. This tranquil state of affairs has been shattered and an army of mathematicians is out there trying to pick up the pieces and assemble them into a more meaningful whole. The twin engines driving this change are modeling and graphical visualization.

The Evolving ODE Course

About a decade ago, serious changes in the ODE course started to take place, mostly because of the emergence of (relatively) inexpensive platforms and the ready availability of powerful, (relatively) easy-to-use software, both numerical solvers and computer algebra systems (CAS).[2] The impact of this computing capability was first felt in the calculus courses, where graphical visualization made it possible to experiment with new modes of instruction. A chronicle of the development of the "new calculus" is contained in other articles of this collection, but from our vantage point the most significant change in that course is its reliance on using computers to solve differential equations which arise in models from the client disciplines. This is hardly a surprising development since differential equations are a window to an unending supply of excellent applications, and calculus is the key to understanding differential equations and the modeling environment in which they arise.

The effect of graphical visualization and modeling on the differential equations course itself has been no less profound. More and more modeling projects requiring computer work are finding their way into the course, mostly at the expense of theory and the derivation of solution formulas. In order to restore some balance, the computer graphics component of the course has been packaged in a variety of ways: sometimes the computer work goes on in a computer laboratory class that meets once a week to work on assigned projects (experiments) from a lab notebook (students can have questions answered when they arrive, but no choice of platform). Other courses treat assigned projects much as homework, letting students use their favorite platform to complete the work (a TA holds regular office hours in a computer lab). Still others find a way to create an environment in which computer graphics and a CAS package completely envelop the course (help features cleverly built into the environment, but no flexibility in the choice of a platform). In any case, the structure of the ODE course seems to be moving in the direction of a hands-on, interdisciplinary approach, but with a variety of modes of implementation. Apparently the "one-size-fits-all" design principle is not at work here.

[2]For an overview of ODE solvers, see the article by Andrew Flint and Ron Wood in the *College Math Journal:* Special Issue on Differential Equations, V. 25, No. 5 (November 1994).

CODEE

In 1992 the NSF/DUE funded a coast-to-coast consortium of seven colleges and universities with the principal aim of enhancing the expertise of mathematics faculty in developing and using interactive computer projects (or "experiments") in differential equations courses.[3] The consortium became known as CODEE (= Consortium for Ordinary Differential Equations Experiments) and at first concentrated on setting up workshops at the consortium institutions to bring together faculty who teach ODEs as a way of both communicating and sharing information and facilitating the hands-on creation of individual experiments. The large number of applicants for the workshops (roughly twice the number that could be accommodated) for the 175 available slots confirmed our belief that many faculty had already started down the road (at least mentally) of using modeling and computer visualization projects in conjunction with their ODE courses. Thus, our experience in differential equations has been somewhat different from that in the reform calculus movement in that we were able to skip over the development stage and go right into the dissemination stage. CODEE also started a newsletter as a way of sharing information with faculty who could not attend a workshop. With eight issues published to date and a list of over 1300 subscribers, we feel that the CODEE newsletter is filling a real need.[4] CODEE now has a home page on the World Wide Web which has all back issues of the newsletter, reviews of ODE solvers, and other pertinent information. WWW service is available with the Universal Resource Locator: `http://www.math.hmc.edu/codee/`

Models and Experiments

The following four examples show some of the features of the new approaches to teaching differential equations. All four are based on phenomena in client disciplines. Students find each of the four models relatively easy to understand, and they learn a lot about the behavior of the solutions of the differential equations involved through the numerical simulations they carry out.

The first example on rise and fall times is old hat (Newton must have known the answers) but is not usually done in the ODE course. The second example on the effectiveness of a communication channel in transmitting a train of pulses was outlined in a CODEE workshop by the electrical engineer Alan Felzer (California State Polytechnic University, Pomona), and reported in the CODEE newsletter (Winter 1993). The third example on the flow of a medication through the body compartments was developed at the same workshop by the mathematician and pharmaceutical consultant Edward Spitznagel (Washington University), and also reported in CODEE (Fall 1992). The autocatalator example is based on work reported by the chemists Peter Gray (Leeds University) and Stephen Scott (Cambridge University) in their book *Chemical Oscillations and Instabilities*, Oxford University Press, 1990. See also *Differential Equations Laboratory Workbook*, R.L. Borrelli, C.S. Coleman, W.E. Boyce; John Wiley & Sons, 1992.

As these models show, numerical solvers are useful whether or not solution formulas are available. Solvers can easily show the effects of changing initial values or parameters (the so-called "sensitivity" of a model to changes in data). This kind of information is essential when studying any model of a physical phenomenon.

Example 1. If a ball is thrown straight up into the air, does it take longer to rise or to fall, or are the times equal? To answer the question we use Newton's Force Laws to write a second-order initial value problem for the ball's motion:

$$mu''(t) = -mg - f(v), \quad u(0) = 0, \; v(0) = v_0$$

where $u(t)$ is the height of the ball at time t, $v(t)$ is the velocity $u'(t)$, m is the mass of the ball, g is the gravitational constant, and $f(v)$ is the damping force of the air on the ball. Since that force opposes motion, the sign of $f(v)$ must be opposite to the sign of v. Two commonly used models for the damping force are the viscous case where $f(v) = \alpha v$ and the Newtonian model where $f(v) = \beta v|v|$, α and β positive constants. The Newtonian model is valid for dense bodies, the viscous model for less dense bodies (e.g., a whiffle ball). After scaling time, position, and velocity to dimensionless variables, the viscous damping model is

$$\frac{d^2x}{ds^2} = -1 - \frac{dx}{ds}, \quad x(0) = 0, \; x'(0) = w_0$$

[3] The consortium member institutions and their representatives currently are Cornell University (J. Hubbard, B. West), Harvey Mudd College (R. Borrelli, C. Coleman), Rensselaer Polytechnic Institute (W. Boyce, W. Siegmann), St. Olaf College (A. Ostebee, M. Richey), Stetson University (M. Branton, M. Hale), Washington State University (M. Kallaher, M. Moody), West Valley College (D. Campbell, W. Ellis).

[4] There is no charge for U.S. subscribers (as long as the NSF funds last). To subscribe, send e-mail to codee@hmc.edu, or write to CODEE, Math Dept., Harvey Mudd College, Claremont, CA 91711.

The Newtonian damping model (with different scale factors but using the same variable names) is

$$\frac{d^2x}{ds^2} = -1 - \frac{dx}{ds}\left|\frac{dx}{ds}\right|, \quad x(0) = 0,\ x'(0) = w_0$$

Figure 1 shows the graphs of the scaled position $x(s)$ (upper) and velocity $y(s) = x'(s)$ (lower) in the viscous case for $w_0 = 1, 2, 4$.

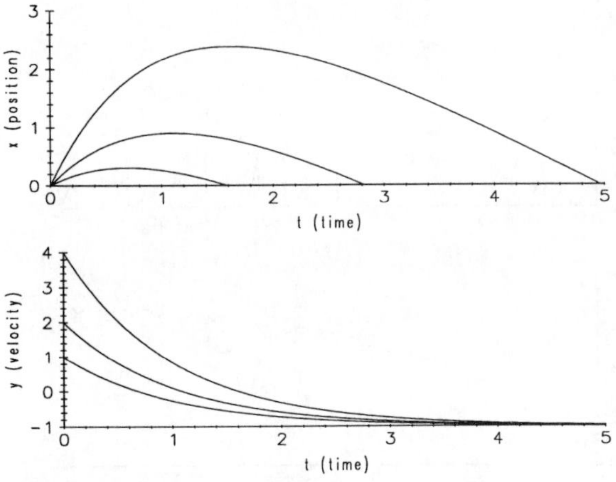

Figure 1: Rise and fall with viscous damping.

Figure 2 shows similar graphs in the Newtonian case for $w_0 = 1, 5, 9$.

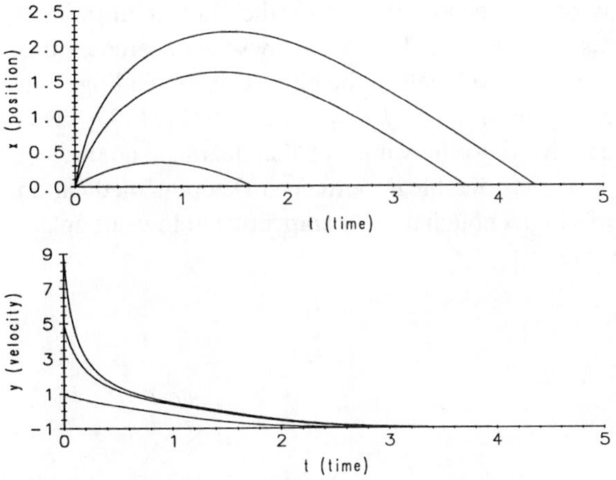

Figure 2: Rise and fall with Newtonian damping.

In both cases, it is a little hard to tell from the low initial velocity graphs whether the rise time to the maximum height is greater than, equal to, or less than the fall time. However, the graphs corresponding to a higher initial velocity clearly suggest that the fall time in both cases is longer than the rise time. Why should it take longer to rise than to fall? Hint: the forces of gravity and friction act in concert on the way up, but oppositely on the way down. The better students could be asked to come up with a mathematical proof (at least in the viscous case). Can you show that the rise time is less than the fall time by examining the concavity of the velocity curve?

Example 2. A bread and butter problem in electrical engineering is to find the band of frequencies that can be transmitted by a simple communications channel and still be recognizable at the receiving end. The channel is modeled by an electrical circuit:

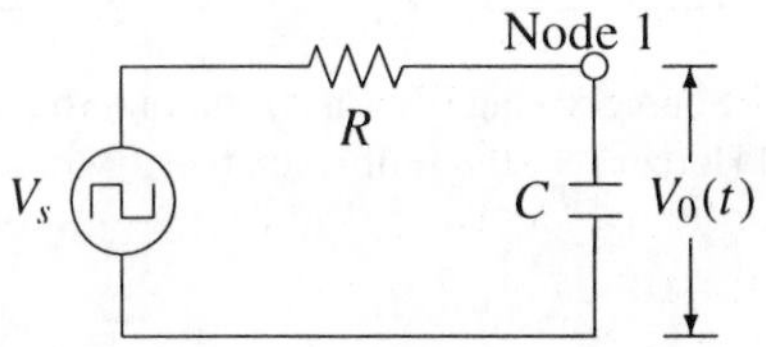

The source voltage $V_s(t)$ is the "message sent," the output voltage $V_0(t)$ is the "message received." The constants R and C are the resistance and capacitance of the transmission lines. From Kirchhoff's Law, the current entering node 1 equals the current leaving:

$$\begin{aligned}\text{current entering} &= \frac{1}{R}[V_s(t) - V_0(t)] \\ &= CV_0'(t) = \text{current leaving}\end{aligned}$$

For a specific example, set $RC = 1$, $V_0(0) = 0$ to obtain the initial value problem:

$$V_0'(t) + V_0(t) = V_s(t), \quad V_0(0) = 0$$

Suppose that the source voltage is an on-off pulse of period T milliseconds (i.e., of frequency $f = 10^3/T$ cycles per second, or Hertz) with amplitude 1. The source voltage is "on" for the first half of each cycle and "off" for the second half. Using a numerical solver to solve the initial value problem for frequencies $f = 1250$ Hertz (Figure 3) and 33 Hertz (Figure 4) we conclude that the communication channel badly attenuates and distorts high frequency signals, but transmits low frequency signals reasonably accurately.

Students can experiment with various values of f and find the highest frequency for which the output signal still resembles the input. The initial value problem for $V_0(t)$ is first-order, linear, and easily solved symbolically:

$$V_0(t) = e^{-t}\int_0^t e^u V_s(u)\,du$$

Students in an introductory differential equations course have difficulty evaluating the integral if V_s is a periodic pulse, but the numerical approach is easy and informative.

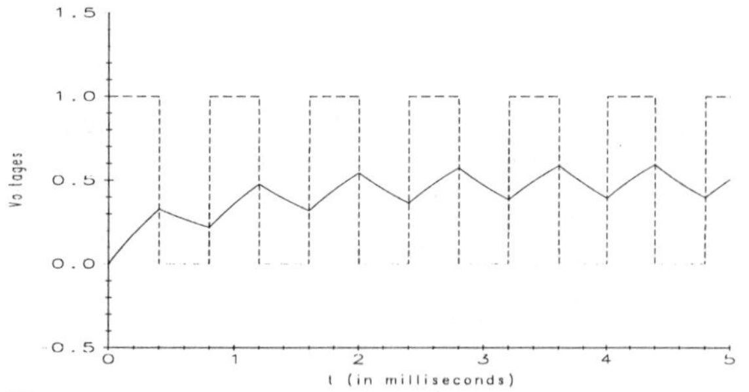

Figure 3: Source voltage (dashed), output voltage (solid); $f = 1250$ Hertz ($T = 0.8$ milliseconds).

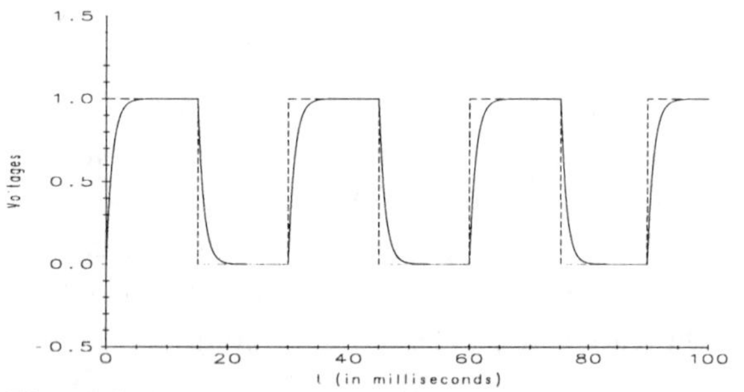

Figure 4: Source voltage (dashed), output voltage (solid); $f = 33$ Hertz ($T = 30$ milliseconds).

Example 3. There are many over-the-counter products for the relief of a runny nose and watering eyes. These typically contain an antihistamine, chlorpheniramine maleate (CPM). CPM passes from the GI tract to the bloodstream and then is cleared from the bloodstream by the liver and kidneys. The rate of passage of CPM from one body compartment to another is proportional to the amount of CPM in the first compartment. Each constant of proportionality k_i is related to the half-life τ_i of CPM in compartment i, $k_i\tau_i = \ln 2$. For the "average individual" the half-life is about one hour in the GI tract and about thirty hours in the bloodstream; the corresponding rate constants are $k_1 = 0.6931$/hour and $k_2 = 0.0231$/hour. The rate equations for the amounts $x(t)$ and $y(t)$ of CPM in the GI tract and bloodstream, respectively, are

$$x' = I(t) - 0.6931x, \quad y' = 0.6931x - 0.0231y$$

where $I(t)$ is the input dosage rate. Figure 5 shows the amounts of CPM in the GI tract (upper graph) and in the bloodstream (lower graph) if a fast-dissolving pill delivers a unit dose of CPM to the GI tract over a half-hour period, and the dose is repeated every six hours for 125 hours.

The rate equations are linear and can be solved one at a time "from the top down," but the solution formulas are messy because of the on-off character of $I(t)$. Using a numerical solver to approximate $x(t)$ and $y(t)$ is clearly the way to go here.

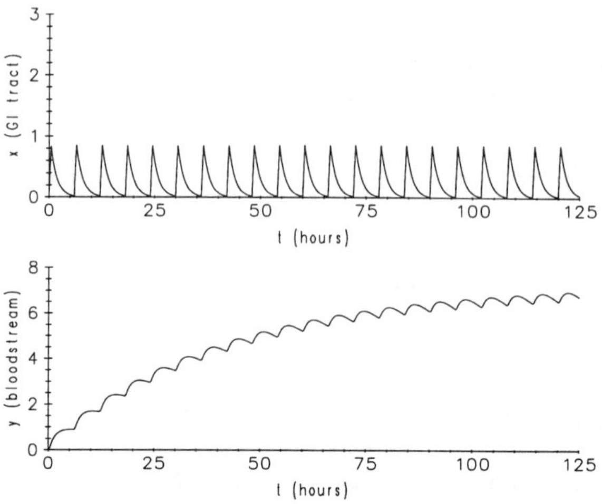

Figure 5: Antihistamine levels in GI tract and bloodstream: one dose every six hours.

The value of the clearance coefficient k_2 varies from person to person. For the young and healthy k_2 may be as large as 0.05, but for the old and infirm may be as small as 0.01. Figure 6 shows the differences this makes in CPM levels in the bloodstream; reading from the top curve down, $k_2 = 0.01, 0.0231, 0.05$. This "sensitivity" to the values of the clearance coefficient k_2 is of considerable importance in keeping medication levels high enough to be therapeutic but low enough to be safe.

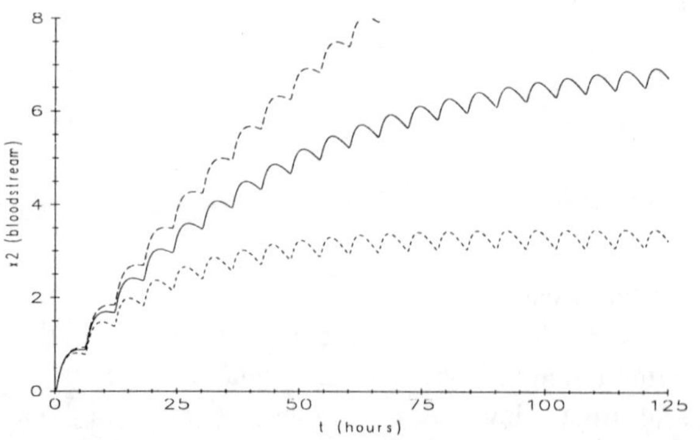

Figure 6: Sensitivity of antihistamine levels in bloodstream to changes in the clearance coefficient.

Example 4. Chemists have recently discovered that concentrations of some chemical species oscillate violently during a chemical reaction. The autocatalator is a model for one of these reactions. Denoting chemical species by X_i, and their concentrations at time t by $x_i(t)$, $i = 1, 2, 3, 4$, the autocatalytic model may be described by the schematic

$$X_1 \xrightarrow{k_1} X_2, \quad X_2 \xrightarrow{k_2} X_3,$$

$$X_2 + 2X_3 \xrightarrow{k_3} 3X_3, \quad X_3 \xrightarrow{k_4} X_4,$$

and mathematically modeled by the system of rate equations in the concentrations

$$\begin{aligned} x_1' &= -k_1 x_1 \\ x_2' &= k_1 x_1 - k_2 x_2 - k_3 x_2 x_3^2 \\ x_3' &= k_2 x_2 - k_4 x_3 + k_3 x_2 x_3^2 \\ x_4' &= k_4 x_3 \end{aligned}$$

with initial conditions $x_1(0) = \alpha$, $x_2(0) = x_3(0) = x_4(0) = 0$. Species X_1 is a reactant that decays to an "intermediate" X_2, X_2 both decays to a second intermediate X_3 and reacts with X_3, producing more of X_3 than is consumed in that autocatalytic process. X_3 decays to the final product X_4. Time, concentrations, and rate constants k_i have been scaled into dimensionless form. Realistic values for the scaled rate constants k_i are $k_1 = 0.002$, $k_2 = 0.08$, $k_3 = 1$, $k_4 = 1$, and the scaled initial value α of x_1 is first taken to be 50. The evolution of x_1, x_2, x_3, and x_4 (x_1 and x_2 are scaled by a factor of 20 to fit into the graph) is shown in Figure 7: x_1 (solid), x_2 (long dashes), x_3 (short dashes), x_4 (long/short dashes). There are no surprises here—this is "classical" chemical kinetics behavior.

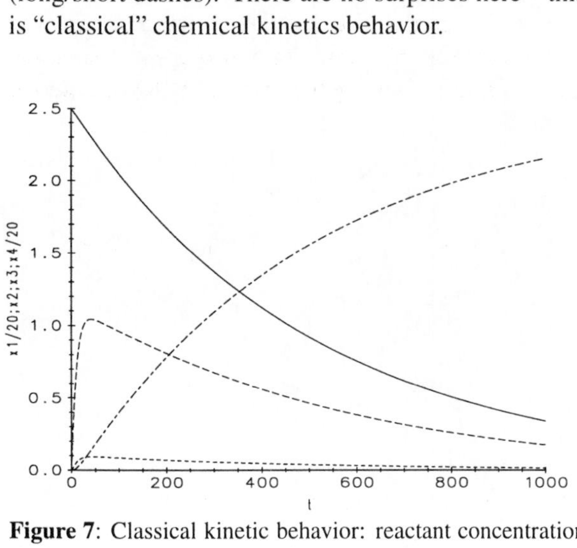

Figure 7: Classical kinetic behavior: reactant concentration declines, intermediates rise and then decline, final product rises: $x_1(0) = 50$.

Now let $\alpha = 500$ and see what happens. Figure 8 shows the surprising oscillations in the concentrations of the two intermediates X_2 and X_3 (plotted are $x_1/200$, x_2, x_3, $x_4/200$).

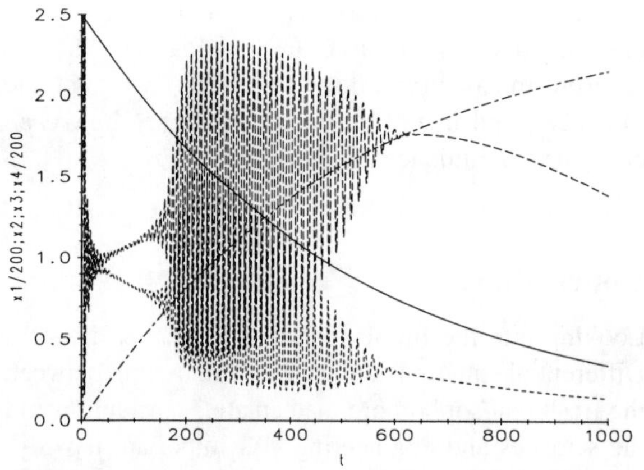

Figure 8: If $x_1(0) = 500$, oscillations in the intermediate concentrations turn on and later turn off.

Apparently when the concentration x_1 of X_1 drops to a certain level (about 400 according to the graph), a bifurcation occurs to large amplitude oscillations in x_2 and x_3. The oscillations stop when x_1 has declined to approximately 150, and "normal" kinetic behavior resumes. Since there are no solution formulas for x_2 and x_3, this "sensitivity" to the initial concentration can only be seen in the laboratory, via computer simulations of the model rate equations, or by mathematical theory (Hopf bifurcation theory is most likely) applied to the model rate equations.

Reform Calculus and the New ODEs

Reform calculus is based on three approaches to each basic concept: algebraic (deriving and using formulas), geometric (drawing and interpreting graphs), and numeric (estimating values of functions). In addition there is a pronounced trend toward the use of practical and everyday phenomena as the starting point for the introduction of a mathematical definition, technique, or theorem. All of this is a reversal of the approaches often used in the recent past: first the mathematical definitions and theorems, then the techniques needed to derive formulas, and finally a few applications (if there is time left in the course). The new differential equations course is moving in the same direction as reform calculus. The four examples given in this note exhibit some of the new approaches, but with an extra twist: Although solution formulas for the ODEs in Examples 1, 2, 3 can be derived, they are not particularly simple, nor are they very helpful in resolving the questions posed in the examples. For the system of ODEs in Example 4, there are no solution formulas. This does *not* mean that solution formulas, where they exist, are to be ignored, but that formulas are not the only way to understand how solutions of ODEs behave.

Reform calculus and the new ODEs will continue to evolve, and it will be interesting to see how each continues to influence the other.

Conclusions

Looking into the future is for the foolhardy, but .... Differential equations will be the primary link between the freshman/sophomore mathematics curriculum and the sciences and engineering. As important new dynamical phenomena are explored in those areas, new models for the phenomena need to be constructed and introduced into the ODE course. Some of these models will last, others won't, but students will see that the differential equations course is "on the frontier." Graphical visualization will become an even more important tool in understanding the behavior of solutions of differential equations. What they see on the computer screen will lead students to make conjectures about solution behavior. The faculty's role at that point will be to discuss the mathematical techniques that will support or refute these conjectures. That is, the differential equations course will become more like experimental science or engineering—but with the fundamental difference that full certainty about the nature of solution behavior must rest on mathematical proofs. Since computer solvers will play such a central role, all of us will become more knowledgeable about numerical methods and software, their strengths, and their pitfalls. To summarize, the introductory differential equations course will rest on three pillars: mathematical theory and formulas, computer visualization and numerics, and models of physical phenomena.

Changing Calculus: Its Impact on the Post-calculus Curriculum

David Carlson and *Wayne Roberts*
SAN DIEGO STATE UNIVERSITY and MACALESTER COLLEGE

To the extent that changes in the calculus sequence achieve their goals, students who take subsequent mathematics courses will have different expectations and, one hopes, abilities closer to what we have always hoped to see in students entering these courses.

Students should understand that clear written expression of ideas is a part of mathematics, and they should be better at exposing these ideas than students who think of mathematics homework in terms of lists of exercises, very similar to worked examples in the text, that call for getting an answer in the back of the book. They should be ready to work on a problem that takes time, that needs to be thought about from several different viewpoints. They should be looking for connections to other areas of mathematics as well as to other disciplines. They should know some of the heuristics of problem solving, and be equipped to employ technological resources intelligently to explore a problem numerically, graphically, or symbolically.

Faculty members who have had good experiences with a new approach to calculus report that there is a direct effect on how they teach advanced courses. They feel that their approach to group work, frequently not a new idea in advanced courses, is more effective. They feel less compelled to cover every topic in class, trusting that students can learn some things from written sources. Recalling the effort in calculus to motivate and present ideas in ways that will actively engage the students, they give much more attention to these same issues in their more advanced courses.

Specific Issues to Anticipate

Skeptics have worried that students coming out of the modern calculus course will have weak computational skills, and proponents have conceded that it might be so. It is certainly an issue to think about. One hopes that it will be considered in a realistic assessment of the past, that it will be remembered that we have been complaining for years that students seem not to know, or to understand, or to remember what we have taught them. Either way, whether it is something new or more of the same familiar problem, it is probably realistic to think that some backfilling may be necessary. What should be new is that students with a better idea of the underlying concepts and an increased ability to read a handout or a book, should be better equipped to learn on their own, or at least with a modest amount of help.

In the same way, it should be anticipated that some topics once included in the calculus course might no longer be there. If your course in applied analysis needs hyperbolic functions, it may be that they need to be introduced as something new. If partial fractions are needed to make use of Laplace transforms or to present the notion of generating functions, they too may have to be taught as something new. For the reasons cited above, presenting these special topics should not cause great inconvenience if the need is anticipated. There is little difference between reviewing a topic commonly forgotten or presenting that same topic when it is needed.

There is also a bright side to this matter of what might be anticipated. If anything, students should be better prepared to write formal proofs than were those students whose previous experience with proof consisted of rote memorization of a proof that they were warned might be on an exam. Certainly the expectation in the modern calculus course that students explain their answer or justify their approach to a problem should set the stage for writing more formal proofs. All should set the stage for writing more formal proofs. All too often, students have felt utterly at sea when asked to make the transition from the drill enforced manipulations of calculus to the verbal emphasis of linear algebra and later courses.

In the same vein, students coming out of a course in which projects were routinely assigned should have a much better idea of what is expected of them when confronted with the problem sets commonly used in advanced courses. The attention to problem-solving heuristics, and the ability afforded by technology to examine greater numbers of examples and more complicated examples, should also result in more student success with these problem sets.

Some Examples

The first author has had extensive experience with building on the ideas of the new calculus to teach a revitalized linear algebra. Ideas once taught by carefully orchestrated theoretic development can now be anticipated by students who can look at many more examples than were possible when every matrix multiplication had to be done by hand. Transformations from $\mathbb{R}^2$ to $\mathbb{R}^2$ are better understood when students can see the geometric effect of minor changes in the matrix representing the transformation. Students can better focus on the meaning of eigenvalues when they don't first get bogged down in the hand computations necessary to find them. Many more examples could be given.

Moreover, students are much more easily convinced of the need for certain theoretical concepts. One need not try many matrix computations to see the value of diagonalization via similarity. The observation that certain calculators never met a matrix they could not invert sends a strong message about the need to control round off errors, and the need to be alert to ways in which machines can go wrong. The need for some theoretical understanding stands out in bold relief.

This procedure of building on computational experience is also an effective way to develop the all important mathematical concept of abstraction. Students don't hate abstraction in and of itself. They hate it when it seems irrelevant, but they recognize its utility when they see that it brings together seemingly disparate concepts that they understand individually. This author's students apply the Gram–Schmidt orthogonalization process to sets of functions. They work with linear maps of the form $L(X) = AX - XC$ acting on square matrices X. Given the coefficient vectors of two polynomials, the students figure out how to put them into a big matrix so that the row reduced echelon form of the big matrix exhibits the coefficient vectors of the quotient and remainder of the original polynomials. These activities are terribly difficult for them; but through them they apply linear algebraic ideas, which they are most comfortable with in the setting of n-tuples, to functions, matrices, and polynomials. This way they are prepared to understand the axioms of a vector space and why we teach them.

The second author recently taught what he believes to be the most successful real analysis course in his thirty years of teaching. Using *A Radical Approach to Real Analysis* [David Bressoud, MAA, 1995], the course began by asking students just out of the calculus sequence to plot a sequence of functions that were actually partial sums of the Fourier series for a step function. Using *Mathematica* (the code was supplied in the text), the students could, in a few minutes, look at accurate plots of sequences with more terms than Fourier probably saw in his lifetime. The convergence of these continuous functions to a discontinuous function was obvious to everyone, and it set the stage for a description of the historic struggle to clarify the concept of a function, to define convergence, to understand continuity, etc. The formal definition of continuity did not come until halfway through the course. When it finally came, no one asked why we needed to worry about such an arcane concept.

Several things were noteworthy about the experience. The instructor was not adept at *Mathematica*, and neither were many of the students. By forming groups, however, we were able to use the experience of those who were at home in the *Mathematica* lab so that there was no lost time while students became familiar with the technology. The groups spent their time answering the text's penetrating questions about the pictures they saw. The students were delving into substantial mathematics that, by term's end, took us much deeper than the author had ever gotten in a single semester course. This was not a history course. Yet the students learned a lot about history, and about the struggle that had gone into creating this mathematics. Motivation was never a problem as the students got caught up in how all these matters would be resolved.

Other examples abound. The course in complex variables can be motivated with non-trivial mathematics following the account given by Sondheimer and Rogerson [*Numbers and Infinity*, New York, Cambridge University Press, 1981] of how early investigators came to realize that there was a practical need for complex numbers. If it is useful for students to see images of regions in the plane as transformed by a conformal mapping (and pages of these pictures contained in such classic texts as the one by Churchill indicate that we have traditionally thought it is useful), then how much more useful is it for students to be able to explore for themselves in half an hour many more such pictures than any publisher can afford to put in a text?

Conclusion

Suffice it to say that we have only begun to explore the implications of the new calculus course for the teaching of advanced courses. The students should be coming into these courses better prepared to see the things we want them to see, perhaps even to see things we have not thought of. It should be easier to accommodate the increasingly diverse audiences we have as the students are taught to be better readers, more effective learners, and more independent thinkers. The challenge for us is to exploit the opportunities that present themselves in this most exciting of times to be teaching mathematics.

A Round-Table Discussion with the Client Disciplines

Sheldon P. Gordon
SUFFOLK COMMUNITY COLLEGE

MATH: I'd like to welcome all of you to this round-table discussion on the impact of the new calculus courses on the client disciplines. The panelists who have taken the time to share their experiences and perspectives with us are: Stephen Boyd (electrical engineering, Stanford University), Tom Daula (economics, US Military Academy), Doyle Daves (chemistry, RPI), David Hanson (chemistry, SUNY/Stony Brook), T. J. Mueller (biology, Harvey Mudd College), John Prados (chemical engineering, NSF), Wayne Roberge (physics, RPI), Mike Ruane (electrical engineering, Boston University), Clifford Schwartz (physics, SUNY/Stony Brook), and Angela Stacy (chemistry, UC, Berkeley).

We in the mathematics community are extremely interested in how the changes we are making in calculus are perceived in your fields. Are these changes appropriate to your needs, or will they cause severe problems down the road?

Let's begin with the question of how your own disciplines have changed over the years, particularly with regard to the mathematical needs of your students. Anybody want to start?

EE: Let me take a stab at that. There have been tremendous changes in engineering over the last 20 years, and these changes are now being reflected in engineering education programs. A lot of the plug-and-chug approach has given way to a much greater emphasis on computing and modeling.

ECO: Modeling—that's the key word in modern economics.

EE: Exactly. We're now freed of the necessity of looking at just those simple cases that admit of closed form solutions. We can look at more realistic models which include more relevant factors.

PHYS: Yes. For example, we always looked at projectile motion with only gravity and then stepped up to include air resistance where the force is proportional either to the square of the velocity or the square root of the velocity. Why? Because those are the cases you can integrate in closed form. But do you really think that either model is the precise truth? In the past, we've been limited by our analytic tools. Now that we have better tools, we in physics are also looking at more realistic situations.

EE: That's exactly the kind of thing I mean. We still start with the fundamental ideas, usually with some form of linearity assumption. However, we quickly move to nonlinear systems with perturbations and then compare the results of linearizing the model to the nonlinear model.

MATH: But what does that mean in terms of your courses, or, for that matter, our courses?

EE: Our students must be able to interpret the behavior of solutions based on graphical output. They must develop a much better understanding of the function concept. More and more, engineering students are looking at numerical and graphical representations; less and less are they looking at symbolic methods. In general, the profession of electrical engineering drives the need for qualitative methods. In fact, the message I continually have to give some of my students is that the vast majority of things do not have algebraic formulas; their calculus training was just too lopsided in emphasizing symbol moving. When they need analytic representations, we expect them to use some sophisticated computer packages.

ChE: That's not only true in electrical engineering, but across all the engineering disciplines. There is far less of the old-fashioned pencil-and-paper mathematical manipulation. In all areas, when students need to evaluate integrals or solve differential equations, they are expected to turn to *Mathematica*, *Maple* or *MathCad*. What is critical is that the students develop a better conceptual understanding of the mathematical ideas and techniques: what is the effect of changing parameters in a model? What is the behavior of a function, particularly one with several parameters? How do you go between the symbolic representation and the graphical representation of a function? Even something as basic as: how do you distinguish between the dependent and independent variable?

EE: Right! It's the understanding and the applications that count. In terms of what many students are learning today, the traditional symbolic manipulation is totally irrelevant.

PHYS: It's exactly the same in physics today. The heavy manipulation is being relegated to Maple or Mathematica. What we want is for students to bring a basic understanding of fundamental concepts of calculus into their physics courses. Right now, they are very good at taking the derivative mechanically, but have little idea of what the derivative tells them.

EE: Let me give an example of that based on the fact that the current in a circuit is the derivative of voltage. A common problem in electrical engineering texts is: here's the graph of the voltage in a circuit; sketch the graph of the current. It's amazing how many students, who have completed a full three semesters of calculus, have no idea where to begin. All they can do is differentiate any conceivable expression, but they don't know what it means. But this problem underlies how an understanding of the mathematics helps in understanding the engineering situation and vice versa.

CHEM: I'm a bit embarrassed to admit that I haven't evaluated an integral in 40 years! That kind of thing is not really what many of us need calculus for in chemistry. But, we typically require three semesters of calculus for students heading toward organic chemistry, particularly to build up the ability to visualize objects in three dimensions. For students heading toward physical chemistry, we need more, definitely a good course in linear algebra and possibly one in differential equations. However, what seems to be important is for students to understand the meaning and application of derivatives and integrals, how to set up a differential equation and interpret the behavior of its solution. Knowing where to look up integrals seems as effective for most of us as knowing how to integrate. If anything, numerical methods also seem more important now than analytical techniques. Most importantly, students need to bring an understanding of the concepts of calculus.

PHYS: The changes I see happening in physics go beyond that kind of thing to the development of a very different learning environment, one that makes the student a far more active participant in the learning process. Research has shown that the traditional lecture course in physics is only minimally effective in stimulating learning for today's students. We are attempting to replace it with different settings based on collaborative learning, cooperative activities, and interactive learning. Some of us call it a workshop environment. It appears to be very effective as an educational approach, but also better prepares students for the workplace. We believe that students are unlikely to find satisfying employment in physics without effective communication, cooperation and leadership skills, none of which is fostered in the passive learning environment.

ChE: Yes! That's exactly the cornerstone of the changes happening across the engineering disciplines today. The practice of engineering focuses much more heavily on some of the softer skills that we have ignored in the past—things like communication, teamwork, and the need to examine ill-posed problems. Engineers do not work on projects individually; they are always part of teams, and the teams typically involve individuals from a number of different disciplines. Also, companies can no longer afford the luxury of graduating engineers basically serving a one- or two-year apprenticeship as they learn how to be practicing engineers. The employers want to hire people who are able to contribute almost from day one.

The teaching of engineering is rapidly changing to reflect this paradigm with a more integrated approach combining engineering, science, and mathematics done in context. We are attempting to develop a far greater degree of teamwork, communication, and broader information about associated fields. For instance, we would like to see students of engineering learn many of their mathematical skills in the context of solving sophisticated problems. This involves a lot of team teaching, for example. There are a number of pilot programs underway to implement these ideas and they are quite successful—at one location, graduation rates for engineering majors have increased by about 50%.

CHEM: Many of the same themes are being discussed in the chemistry community. Mathematical modeling is becoming increasingly important as commercial software becomes widely available. Differential equations play a big role here, especially in kinetics and quantum mechanics. But the new pedagogy will also entail more student involvement and less lecturing. As in engineering, projects in chemistry are done by teams, not by individuals, so it is particularly important that we place more emphasis on collaborative work in our introductory courses. Of course, chemists are conservative by nature, but I expect that they will fall in line when they see the positive results.

BIO: I think that people in all fields are beginning to realize that the traditional lecture approach is great for covering a lot of material quickly, but is even better at putting the students to sleep. It is critical that we get them actively involved in the learning process. When we do that, they will learn far more and it will stick better. In fact, we must refocus our own perspectives to emphasize learning, not teaching. The professor must become more of a facilitator or learning director than the leader. And that's not easy.

PHYS: But, to do any of that means that we must accept the principle of covering less material, but looking at the core topics in more detail and in greater depth. There are some wonderful traditional texts out there, but they are far too encyclopedic to serve as introductory textbooks. You have to go through them so quickly that most students come out with very little understanding of the fundamental concepts of physics. The key is to focus on how much you learn, not how much you cover.

BIO: I couldn't agree more. We should be worried much more about concept than about content. It's not so terrible if we cover less provided that our students learn how to learn.

PHYS: There is another aspect to our philosophy of having students learn physics in a workshop atmosphere. We find that the application of the concepts immediately after they have been introduced forces the students to confront weak spots in their comprehension and ask the instructor for help. It gets very hard to simply hide in the back of the lecture hall.

MATH: I find it amazing how parallel the thinking appears to be in all of your fields to what we have been saying and doing in improving the mathematics curriculum. Are there any other changes taking place in your fields that relate to mathematics?

BIO: More and more, we are looking at using simulations of biological processes and systems. For instance, they might be on population growth or physiological processes or environmental issues. The advantage to using simulations instead of the actual processes is that they are faster, safer, and far less expensive. When the students get involved in using simulations, we want to change the focus and have them begin to question how appropriate the model is: does it truly model the situation under study? What happens when you vary the parameters in the model? We want them to question the basic assumptions in the model, if it seems appropriate, and to change the model to build a more sophisticated one.

But, when they get into issues such as that, they must come to a deep understanding simultaneously of the biological processes and the modeling process, and the latter requires understanding the mathematics used in developing the model. Typically, that is a differential equations model. Also, we are often looking at discrete processes, so there is a considerable emphasis on difference equations. Many models require a knowledge of linear algebra and certainly some understanding of basic applied statistics.

ECO: The same is true in economics. In fact, the majority of the models we use are discrete, typically based on difference equations. They also tend to involve a stochastic component, so that students need an introduction to probabilistic reasoning early in their studies.

ChE: In the traditional educational approach, we in engineering tended to begin with very general, abstract principles and eventually worked our way down to specific applications. We now realize that this is not the best approach for most students—they are better served by starting with a series of down-to-earth examples and then generalizing to discover the fundamental principles. For instance, we might think of this in the framework of mathematical modeling—look at the behavior of a real system and then abstract the general rules. The key in this process is finding the limitations of the model. For example, electronic components are usually rated according to a range of values under which they behave linearly; but what are the limits under which the linearity assumption breaks down, and how do the components behave outside that interval?

ECO: That sounds remarkably similar to what is happening in modern economics and finance. Our primary objective is to develop mathematical models, but the key is in understanding the assumptions on which the model is based. Questions that students must come to grips with are the sensitivity of the results to changes in the underlying assumptions, what the limitations are of the models, and identifying conditions under which certain types of behavior are possible?

BIO: What I see is the need for students to be prepared to face unknown problems. Too often, students, even very good students, approach every new situation with fear because their previous experience has been very narrow. I'd love to see them willing to just dive in and try to formulate a differential equation to describe a process, and then examine the behavior of the solution to see whether it reasonably describes that process.

MATH: But how do you see getting to that stage?

ChE: One of the major trends we see in engineering is the development of multimedia courseware. That has the potential to do some truly wonderful things in that direction.

BIO: Yes. Computer simulations are an extraordinary tool for involving students in a problem-solving environment. It encourages them to interact at a much deeper level of involvement. Perhaps more importantly, it opens up doorways to them. A textbook approach is very narrowly focused—the author directs the reader along the prescribed course in a totally linear fashion. A lecture approach tends to do more of the same. The teacher leads the discussion, highlighting the points he or she thinks are important, and diverting the students away from questions that they may feel are significant; often this occurs because the instructor may feel ignorant of the side issues. But a true multimedia environment allows the students to go off the primary path to find the answers to the often unexpected questions that arise in their own minds.

PHYS: That's exactly the kind of thing we are doing in physics as well. It allows us to get into real-world issues that you can't get into with purely analytic methods. Good computer courseware provides an interface between the students and physical experiments by collecting data, analyzing the data, allowing the students to visualize the behavior of the variables, and looking at the effects of changes of parameters. As an example, someone already mentioned how current is the derivative of voltage. But any measurement of the voltage includes a certain amount of "noise;" when you differentiate that quantity, the noise gets amplified. Students must understand this as a general principle, and that can best be conveyed to them via computer simulations to see the effects of different assumptions about the noise.

ChE: Another interesting thing we have observed is that many females tend to come into engineering classes with a less well-developed ability for three-dimensional visualization. We have found that the use of graphical technology is very effective as a means for remediating this problem.

MATH: To what extent are the ideas you have been discussing actually being implemented in your fields?

EE: Obviously, there is a broad spectrum of people in every discipline. As someone mentioned above, chemists tend to be conservative. There are conservatives in all fields, except possibly in math where you have made such extensive changes across the discipline. In electrical engineering, most of us are moving in the directions I mentioned before, though I might be considered somewhat of an outlier. Nevertheless, electrical engineering is changing rapidly and the mathematics you teach is clearly there to reinforce the engineering ideas.

ChE: The need to reform engineering is becoming very widely accepted and momentum for change is truly moving rapidly. The NSF has funded the creation of a group of what are called "engineering coalitions" which are designed to develop and implement engineering programs based on the comments we have made. The institutions involved in these coalitions enroll approximately one-third of all engineering students in the country. Further, the math departments at virtually all of these coalition schools are heavily into calculus reform activities, and we have not heard any negative comments from the engineering and related departments at any of them. In fact, ABET (Accreditation Board for Engineering and Technology) visitation teams have typically reported very favorably on the calculus reform activities. Also, when you speak to many of the engineering faculty, a theme that comes through regularly is that they are pleased to hear that the mathematicians are also concerned with their students.

ECO: In economics, we seem to be heading toward a two-tier system. The best schools have moved very far in the directions I discussed above; the focus in their courses is highly mathematical with tremendous emphasis on modeling situations. Many other schools are still giving relatively non-mathematical approaches to the field, and there is a great potential that the students coming out of such programs will find themselves locked out of the job market, at least in terms of the best jobs, and they will find themselves locked out of top graduate schools. But because the world is becoming ever more difficult, and the economic problems we face become comparably more complicated, we will see pressures to upgrade all our offerings. Also, remember that there are many more economics majors than there are engineering majors, so there are some major implications for the mathematics community.

PHYS: There are a number of projects designed to implement these approaches to introductory physics, and their preliminary results have been very positive. Many other institutions are looking into these projects with the intent of modeling them, so we think that these ideas will spread very quickly.

MATH: Most of what you have been saying about the curricular activities in your different fields is very new to me as a mathematician. I am not really aware of the extensive changes taking place in these areas. Is the reverse true? Are you aware of what has been happening in mathematics in terms of changing the calculus and related curriculum?

CHEM: The mathematicians have kept calculus reform a secret. They need to come talk to us about the interrelationships between what they are doing in calculus and our courses. Only a few months ago, someone mentioned to me that we should be making use of the graphing calculators that all calculus students purchased and learned how to use. But we were totally unaware of this. Mathematicians seem to be isolated and not concerned about the rest of the world. There is a big difference if I go to a mathematician and initiate a conversation about incorporating math into chemistry compared to if that person comes to me.

EE: I agree completely. Engineers in general do not know about calculus reform activities. For example, I learned about what was happening because my own son took one of the reform courses; if not for that, I would probably still be in the dark.

MATH: Then what do you suggest that we do?

CHEM: Come over and talk to us. From our discussions here, it is obvious that there is a tremendous degree of commonality of philosophy; let's share it for the good of all.

ChE: That's fine on a local scale. However, the mathematicians should be more active globally at informing us of what is happening in calculus. We in engineering have a variety of publications devoted to engineering education in each of the subspecialties; write some articles describing your activities, your philosophies, and your goals in calculus reform. They will be very welcome. You might also want to look at them yourselves to get a better feel for what is happening in the reform projects in engineering.

PHYS: I agree completely. There is a journal in physics also dedicated to education and articles describing the changes in calculus would be extremely welcome. Also, come to some of our conferences and give presentations; show us what you are doing.

EE: You should also come to some of the national engineering education meetings. You will be very welcome there. And maybe you should invite us to come to your meetings, both to let you know what we are doing and to let us find out more detail about what you are doing.

MATH: Thank you all for your comments and insight. It has been a very informative and eye-opening experience. I hope that the discussion we have conducted here will be the start of an on-going dialogue among all our disciplines.

Acknowledgment. The above discussion was culled by the author from a series of individual interviews with and talks by the individuals listed at the beginning of this article. The author is extremely appreciative of the time and cooperation extended by these colleagues, and wishes to thank them for their interest and concern for the education of their own students, and their willingness to assist us in the mathematics community.

Calculus for a New Century

Under this heading, the call went forth for the reform of the way calculus was being taught in this country. The goal was clear; we were to make the course lean and lively. Even a slogan was provided. "Calculus should be a pump, not a filter."

Whether the course will become a pump in the nation's pipeline of scientifically trained personnel remains to be seen. What is clear is that the National Science Foundation became a spigot, and the mathematical community became lively.

Individuals, groups of colleagues in some departments, and larger groups from consortiums of colleges and universities provided leadership. Special conferences and special sessions at regular conferences sprang up. Resources were developed, specifically aimed at calculus teachers or would-be reformers. Technology was brought into play. New channels of communication were opened through computer networks, *UME Trends*, and the circulation of experimental materials.

There was, in fact, an explosion of new materials: pamphlets exhorted students to explore mathematical questions with their graphing calculators; manuals gave instructions for experiments to perform in the computer lab. Some people wrote of modern applications while others tried to develop a sense of the historic roots of the subject. Altogether new textbooks began to appear on the market.

The movement interacted with and in some cases was given real direction by those doing research into how students best learn mathematics. It became clear that whatever changes might be made in the content, the new courses were more likely characterized by changes in pedagogy: group participation in place of some classroom lecturing, collaborative rather than individual homework assignments, learning by discovery, learning through work on open-ended projects, learning to read mathematics independently , and writing to learn.

The new materials did not portend well for making the course lean. Had we but looked at ourselves in the beginning, we might have realized that it would be easier to become lively than to become lean. In any case, it is the new life that has been breathed into the old subject that is the most impressive aspect of the reform effort, and the one most important to carry into the new century.

Research faculties, many of whom had not participated in the teaching of calculus for years, have begun to talk about what should be taught in the introductory course. Their constructive involvement is needed, and one hopes to see it sustained. Faculties at all levels have been reminded that it is our success in teaching this course to students from the client disciplines that determines, in large measure, how colleagues from other disciplines view our departments, and it is our success in teaching this course to all introductory students that determines how students view our discipline.

Efforts to change the teaching of calculus have inevitably raised articulation issues between two-year and four-year institutions, and between all secondary and collegiate institutions. Virtually every such discussion comes sooner or later, usually sooner, to a discussion of the Advanced Placement examinations, and whether they facilitate or impede efforts of secondary teachers to emphasize ideas as opposed to the skills emphasized on the exam. While one could, at times, wish for fewer inflection points in these conversations, it is important that they be continuous.

We are asked whether the movement has succeeded, whether the course for a new century has been created. These are two questions, of course, and the second is easily answered. No. A course has not been created. Several variations of a course have been created, or perhaps more accurately, are being created. No one seems likely to emerge as the new monolithic calculus course, and that point is the key to answering the first question.

The movement has succeeded in getting the mathematics community to think about the course. Recent surveys indicate that about one-third of all students who now take calculus are taking a course that has been affected in some way by the changes of the last ten years. It is probably safe to say that many of those who take a traditional course are in a course that has been planned more thoughtfully and taught with more energy than has been the case in the recent past.

It is no longer assumed that we all know what a calculus course should be, that we know just how it should be taught, and that those who don't find it interesting or intelligible are invariably the ones at fault. The success of the movement is seen from the variety of approaches now being used to teach calculus, and from the probability that we will never settle back to a single approach to teaching so dynamic a subject.

The effect of changes in the calculus on client disciplines has, in many instances, gone from concern to emulation. Incredible though it may seem from inside the mathematical edifice, there are any number of colleagues in the sciences and engineering who see us as having our act together in a way that does indeed allow us to look forward to the new century with confidence. It may very well be that the National Science Foundation is correct in proposing *Mathematical Sciences and Their Applications Throughout the Curriculum* as the next natural step.

It seems clear that the discussions engendered inside and outside of our discipline guarantee that we are poised to offer a calculus course that is, if not lean, at least toned up for action, and manifestly capable of stimulating that lively interaction which must be the essence of any discipline that is to provide intellectual leadership for a new century.

Resources*

Compiled by David A. Smith
Duke University

Calculus Reform

Bookman, Jack, and Charles P. Friedman, "A Comparison of the Problem Solving Performance of Students in Lab Based and Traditional Calculus," pp. 101–116 in *Research in Collegiate Mathematics Education I* (E. Dubinsky, *et al.*, eds.), CBMS Issues in Mathematics Education, Vol. 4, AMS–MAA, 1994.

Culotta, Elizabeth, "The Calculus of Education Reform," *Science*, 255, 1992, 1060–1062.

Douglas, Ronald G., ed., *Toward a Lean and Lively Calculus*, MAA Notes No. 6, 1986.

Douglas, Ronald G., "The First Decade of Calculus Reform," *UME Trends*, 6 (6), January 1995, 1–2.

Flashman, Martin, "Editorial: A Sensible Calculus," *The UMAP Journal*, 11 (1990), 93–95.

Goodman, "Toward a Pump, Not a Filter," *Mosaic*, 22 (Summer 1991), 12–21.

Reed, Michael C., "Mainstreaming Calculus Reform," *SIAM News*, 27 (7), Aug./Sept. 1994, 7.

Schoenfeld, Alan H., "A Brief Biography of Calculus Reform," *UME Trends*, 6 (6), January 1995, 3–5.

Smith, David A., "Trends in Calculus Reform," pp. 3–13 in *Preparing for a New Calculus* (A. Solow, ed.), MAA Notes No. 36, 1994.

Solow, Anita E., ed., *Preparing for a New Calculus*, MAA Notes No. 36, 1994.

Steen, Lynn A., ed., *Calculus for a New Century: A Pump, Not a Filter*, MAA Notes No. 8, 1987.

Tucker, A.C., and J.R.C. Leitzel, eds., *Assessing Calculus Reform Efforts*, MAA Report No. 6, 1995.

Tucker, Thomas W., ed., *Priming the Calculus Pump: Innovations and Resources*, MAA Notes No. 17, 1990.

UME Trends, 6 (6), January 1995, issue devoted to calculus reform.

Washington Center News, Winter 1993, issue devoted to calculus reform, The Evergreen State College, Olympia, WA.

Cooperative and Collaborative Learning; Learning Communities

Bruffee, Kenneth A., *Collaborative Learning: Higher Education, Interdependence, and the Authority of Knowledge*, Johns Hopkins University Press, 1993.

Davidson, Neil (ed.), *Cooperative Learning in Mathematics: A Handbook for Teachers*, Addison-Wesley, 1990.

Davidson, N., and T. Worsham (eds.), *Enhancing Thinking through Cooperative Learning*, Teachers College Press, 1992.

Finkel, D.L., and G.S. Monk, "Teachers and Learning Groups: Dissolution of the Atlas Complex," pp. 83–97 in *Learning in Groups*, Jossey–Bass, 1983.

Gabelnick, F., J. MacGregor, R.S. Matthews, and B.L. Smith, *Learning Communities: Creating Connections Among Students, Faculty, and Disciplines*, Jossey–Bass, 1990.

Goodsell, Anne, *et al.* (eds.), *Collaborative Learning: A Sourcebook for Higher Education*, National Center on Postsecondary Teaching, Learning, and Assessment, 1992.

Hagelgans, N.L., *et al.*, *A Practical Guide to Cooperative Learning in Collegiate Mathematics*, MAA Notes No. 37, 1995.

*Adapted with permission from the *Project CALC Instructor's Guide* by D.A. Smith and L.C. Moore, (c) 1996, D.C. Heath and Co.

Heller, Patricia, Ronald Keith, and Scott Anderson, "Teaching Problem Solving Through Cooperative Grouping. Part 1: Group Versus Individual Problem Solving." *American Journal of Physics*, 60 (7), July 1992, 627–636.

Heller, Patricia, and Mark Hollabaugh, "Teaching Problem Solving Through Cooperative Grouping. Part 2: Designing Problems and Structuring Groups." *American Journal of Physics*, 60 (7), July 1992, 637–644.

Johnson, D.W., R.T. Johnson, and K.A. Smith, *Cooperative Learning: Increasing College Faculty Instructional Productivity*, ASHE–ERIC Higher Education Report No. 4, The George Washington University, 1991.

Kadel, Stephanie, and Julia Keehner (eds.), *Collaborative Learning: A Source Book for Higher Education, Vol. II*, National Center on Postsecondary Teaching, Learning, and Assessment, 1994.

MacGregor, J., "Collaborative Learning: Shared Inquiry as a Process of Reform," pp. 19–30 in *The Changing Face of College Teaching* (M.D. Svinicki, ed.), Jossey–Bass, 1990.

Tips for Teachers: Twenty Ways to Make Lectures More Participatory, Derek Bok Center for Teaching and Learning, Harvard University, 1994.

Weissglass, J., "Small-Group Learning," *American Mathematical Monthly*, 100 (7), Aug.–Sept. 1993, 662–668.

What Works, Vol. I: Building Natural Science Communities: A Plan for Strengthening Undergraduate Science and Mathematics, Project Kaleidoscope, 1991.

The State of Mathematics Education

Cipra, Barry A., "Calculus: Crisis Looms in Mathematics' Future," *Science*, 239, 1988, 1491–1492.

Leitzel, J.R.C. (ed.), *A Call for Change: Recommendations for the Mathematical Preparation of Teachers of Mathematics*, MAA Report, 1991.

Madison, B.L., and T.A. Hart, *A Challenge of Numbers: People in the Mathematical Sciences*, National Academy Press, 1990.

Mathematical Sciences Education Board, *Counting on You: Actions Supporting Mathematics Teaching Standards*, National Academy Press, 1991.

Mathematical Sciences Education Board, *Measuring Up: Prototypes for Mathematics Assessment*, National Academy Press, 1993.

National Council of Teachers of Mathematics, *Assessment Standards for School Mathematics*, NCTM, 1995.

National Council of Teachers of Mathematics, *Curriculum and Evaluation Standards for School Mathematics*, NCTM, 1989.

National Research Council, *Everybody Counts: A Report to the Nation of the Future of Mathematics Education*, National Academy Press, 1989.

National Research Council, *Moving Beyond Myths: Revitalizing Undergraduate Mathematics*, National Academy Press, 1991.

National Research Council, *Renewing U.S. Mathematics: A Plan for the 1990's*, National Academy Press, 1990.

Schwartz, J.L., "The Intellectual Costs of Secrecy in Mathematics Assessment," pp. 132–141 in *Expanding Student Assessment* (Vito Perrone, ed.), ASCD, 1991.

Sigma Xi, *Entry-Level Undergraduate Courses in Science, Mathematics, and Engineering: An Investment in Human Resources*, 1990.

Steen, Lynn A. (ed.), *Heeding the Call for Change: Suggestions for Curricular Action*, MAA Notes No. 22, 1992.

Teaching, Learning, and Assessment: General

Alverno Magazine, May 1992, issue devoted to assessment and outcome-based learning, Alverno College, Milwaukee, WI.

Astin, Alexander W., *Assessment for Excellence*, Oryx Press, 1991.

Belenky, M.F., *et al.*, *Women's Ways of Knowing*, Basic Books, 1986.

Bonwell, C.C., and J.A. Eison, *Active Learning: Creating Excitement in the Classroom*, ASHE–ERIC Higher Education Report No. 1, The George Washington University, 1991.

"Bridging the Gap Between Education Research and College Teaching," NCRIPTAL Accent Series No. 9, 1990.

Davis, James R., *Better Teaching, More Learning*, Oryx Press, 1993.

Davis, Todd M., and Patricia H. Murrell, *Turning Teaching into Learning*, ASHE–ERIC Higher Education Report 93–8, The George Washington University, 1993.

Finster, David, "Applying Development Theory May Improve Teaching," *Wittenberg Today*, Wittenberg College, 1987.

Glasersfeld, Ernst von, "Cognition, Construction of Knowledge, and Teaching," *Synthèse*, 80 (1989), 121–140.

Halloun, I.A., and David Hestenes, "The Initial Knowledge State of College Physics Students," *American Journal of Physics*, 53 (1985), 1043–1055.

Katz, Joseph, and Mildred Henry, *Turning Professors into Teachers*, Oryx Press, 1988.

Kozma, Robert B., and Jerome Johnston, "The Technological Revolution Comes to the Classroom," *Change*, 23 (Jan./Feb. 1991), 10–23.

On Teaching and Learning (occasional collections of articles), Derek Bok Center for Teaching and Learning, Harvard University, 4 vols., 1985, 1987, 1989, 1992.

"Personal Growth as a Faculty Goal for Students," NCRIPTAL Accent Series No. 10, 1990.

Rando, William C., and Lisa Lenze (eds.), *Learning from Students: Early Term Student Feedback in Higher Education*, National Center on Postsecondary Teaching, Learning, and Assessment, 1994.

Reedy, George, "I Am Not At All Convinced That We Professors 'Educate' Students. What We Do Is Force Them To Use Their Minds." *The Chronicle of Higher Education*, Dec. 19, 1990, p. B5.

Steen, Lynn A., "Out from Underachievement," *Issues in Science and Technology*, Fall 1988, 88–93.

Steen, Lynn A., "Reaching for Science Literacy," *Change*, (July/August 1991), 11–19.

Taylor, Kathe, and Bill Moore, "Who's Making the Meaning in the Classroom: Implications of the Perry Scheme," POD Conference paper, 1984.

"Teaching Thinking in College," NCRIPTAL Accent Series No. 7, 1990.

Tobias, Sheila, *Revitalizing Undergraduate Science: Why Some Things Work and Most Don't*, Research Corporation, 1992.

Tobias, Sheila, *They're Not Dumb, They're Different: Stalking the Second Tier*, Research Corporation, 1990.

Weingartner, Rudolph H., *Undergraduate Education: Goals and Myths*, Oryx Press, 1991.

"What Are Academic Administrators Doing to Improve Undergraduate Education?" NCRIPTAL Accent Series No. 8, 1990.

What Works, Vol. II: Resources for Reform, Project Kaleidoscope, 1992.

Teaching, Learning, and Assessment: Mathematics

Ball, Deborah L., "Unlearning to Teach Mathematics," *For the Learning of Mathematics*, 8 (1988), 40–48.

Benezet, L.P., "The Teaching of Arithmetic I, II, III: The Story of an Experiment," *Humanistic Mathematics Newsletter*, #6, May 1991, pp. 2–14 (reprinted from *The Journal of the National Education Association*, Nov. 1935, Dec. 1935, Jan. 1936).

Borasi, Rafaella, "The Invisible Hand Operating in Mathematics Instruction: Students' Conceptions and Expectations," *NCTM Yearbook 1990*, pp. 174–181.

Bullock, Richard, and Richard Millman, "Mathematicians' Concepts of Audience in Mathematics Textbook Writing," *PRIMUS*, 2 (1992), 335–347.

Case, Bettye Anne (ed.), *You're the Professor, What Next? Ideas and Resources for Preparing College Teachers*, MAA Notes No. 35, 1994.

Cipra, Barry A., "Untying the Mind's Knot," in *Heeding the Call for Change* (L.A. Steen, ed.), MAA Notes No. 22, 1992, pp. 163–181.

Cobb, George, "Teaching Statistics: More Data, Less Lecturing," *UME Trends*, October 1991, pp. 3, 7.

Dubinsky, E., A.H. Schoenfeld, and J.J. Kaput (eds.), *Research in Collegiate Mathematics Education I*, CBMS Issues in Mathematics Education, Vol. 4, AMS–MAA, 1994.

Farmer, Tom, and Fred Gass, "Physical Demonstrations in the Calculus Classroom," *The College Mathematics Journal*, 23, 1992, 146–148.

Harel, G., and E. Dubinsky (eds.), *The Concept of Function: Aspects of Epistemology and Pedagogy*, MAA Notes No. 25, 1992.

Kaput, J.J., and E. Dubinsky (eds.), *Research Issues in Undergraduate Mathematics Learning: Preliminary Analyses and Results*, MAA Notes No. 33, 1994.

Kulm, Gerald (ed.), *Assessing Higher Order Thinking in Mathematics*, American Association for the Advancement of Science, 1990.

Mitchell, Richard, "The Preconception-Based Learning Cycle: An Alternative to the Traditional Lecture Method of Instruction," *PRIMUS*, 2 (1992), 317–334.

Monk, G.S., "Students' Understanding of Functions in Calculus Courses," *Humanistic Mathematics Network Newsletter*, No. 2, March 1988.

Papert, Seymour, *Mindstorms: Children, Computers, and Powerful Ideas*, Basic Books, 1980.

Pólya, George, "Reprints of Papers on Teaching and Learning in Mathematics," pages 473–603 in *George Pólya: Collected Papers, Vol. IV*, edited by Gian-Carlo Rota, MIT Press, 1984.

Selden, Annie, and John Selden, "Constructivism in Mathematics Education: A View of How People Learn," *UME Trends*, March 1990, p. 8.

Selden, John, Alice Mason, and Annie Selden, "Can Average Calculus Students Solve Nonroutine Problems?" *Journal of Mathematical Behavior* 8 (1989), 45–50.

Selden, John, Annie Selden, and Alice Mason, "Even Good Calculus Students Can't Solve Nonroutine Problems," preprint, 1990.

Schoenfeld, Alan H., "When Good Teaching Leads to Bad Results: The Disasters of 'Well-Taught' Mathematics Courses," *Educational Psychologist* 23 (1988), 145–166.

Schoenfeld, Alan H., ed., *Mathematical Thinking and Problem Solving*, Lawrence Erlbaum, 1994.

Schoenfeld, Alan H., *et al.*, *Student Assessment in Calculus*, Report of the NSF Working Group on Assessment in Calculus, National Science Foundation, 1996 (to appear).

Smith, D.A., and J. Bookman, "Assessment in a Technological Age," *Proceedings of the Seventh Annual International Conference on Technology in Collegiate Mathematics*, November 17–20, 1994, Addison-Wesley, to appear 1995.

Zimmermann, Walter, and Steve Cunningham, eds., *Visualization in Teaching and Learning Mathematics*, MAA Notes No. 19, 1991.

Technology in Teaching and Learning

"The Computer Revolution in Teaching," NCRIPTAL Accent Series No. 5, 1989.

Johnston, J., and S. Gardner, *The Electronic Classroom in Higher Education: A Case for Change*, NCRIPTAL, 1989.

Karian, Zaven A. (ed.), *Symbolic Computation in Undergraduate Mathematics Education*, MAA Notes No. 24, 1992.

Leinbach, L.C., *et al.* (eds.), *The Laboratory Approach to Teaching Calculus*, MAA Notes No. 20, 1991.

Moore, L.C., and D.A. Smith, "Project CALC: Calculus as a Laboratory Course," pp. 16–20 in *Computer Assisted Learning*, Lecture Notes in Computer Science 602, 1992.

Smith, David A., *et al.* (eds.), *Computers and Mathematics: The Use of Computers in Undergraduate Instruction*, MAA Notes No. 9, 1988.

Stroyan, Keith, *et al.*, "Computers in Calculus Reform," *UME Trends*, 6 (6), January 1995, 14–15, 31.

Zimmerman, W., and R.S. Cunningham, *Visualization in Teaching and Learning Mathematics*, MAA Notes No. 19, 1991.

Writing as a Teaching and Learning Tool

Bell, Elizabeth, and Ronald Bell, "Writing and Mathematical Problem Solving: Arguments in Favor of Synthesis," *School Science and Mathematics*, 85 (1985), 210–221.

Connolly, Paul, and Teresa Vilardi (eds.), *Writing to Learn Mathematics and Science*, Teachers College Press, 1989.

Gopen, G.D., and D.A. Smith, "What's an Assignment Like You Doing in a Course Like This? Writing to Learn Mathematics," *The College Mathematics Journal*, 21 (1990), 2–19.

Sterrett, Andrew (ed.), *Using Writing to Teach Mathematics*, MAA Notes No. 16, 1990.

Turner, Judith A., "Math Professors Turn to Writing to Help Students Master Concepts of Calculus and Combinatorics," *The Chronicle of Higher Education*, Feb. 15, 1989, A1, A14.

Problems and Projects (Sources)

Cohen, Marcus, *et al.*, *Student Research Projects in Calculus*, MAA Spectrum Series, 1991.

Dudley, Underwood (ed.), *Resources for Calculus Collection Vol. 5: Readings for Calculus*, MAA Notes No. 31, 1993.

Fraga, Robert (ed.), *Resources for Calculus Collection Vol. 2: Calculus Problems for a New Century*, MAA Notes No. 28, 1993.

Hilbert, Stephen, *et al.*, *Calculus: An Active Approach with Projects*, Wiley, 1994.

Jackson, M.B., and J.R. Ramsay (eds.), *Resources for Calculus Collection Vol. 4: Problems for Student Investigation*, MAA Notes No. 30, 1993.

Solow, Anita E. (ed.), *Resources for Calculus Collection Vol. 1: Learning by Discovery*, MAA Notes No. 27, 1993.

Straffin, Philip (ed.), *Resources for Calculus Collection Vol. 3: Applications of Calculus*, MAA Notes No. 29, 1993.

Periodicals

About Teaching, The Center for Teaching Effectiveness, University of Delaware.

AMATYC Review (The), American Mathematical Association of Two-Year Colleges, Memphis, TN.

American Mathematical Monthly (The), Mathematical Association of America, Washington, DC.

College Mathematics Journal (The), Mathematical Association of America, Washington, DC.

Computer Algebra Systems in Education Newsletter, U.S. Military Academy, West Point, NY, 1987–.

Cooperative Learning and College Teaching, New Forums Press, Stillwater, OK, 1991–.

CUE Newsletter, Collaboration in Undergraduate Education, La Guardia Community College, NY, 1994–.

Educational Studies in Mathematics, D. Reidel Publishing Co., Dordrecht, Holland.

Eightysomething! Texas Instruments, Dallas, TX.

Focus on Calculus, John Wiley & Sons.

Focus on Learning Problems in Mathematics, Center for the Teaching and Learning of Mathematics, Framingham, MA.

Focus, The Newsletter of the MAA, Mathematical Association of America, Washington, DC.

For the Learning of Mathematics, FLM Publishing, Montreal, Canada.

International Journal of Mathematical Education in Science and Technology, Taylor & Francis Ltd., England.

Journal for Research in Mathematics Education, National Council of Teachers of Mathematics, Reston, VA.

Journal of Computers in Mathematics and Science Teaching, Association for the Advancement of Computing in Education, Charlottesville, VA.

Journal of Mathematical Behavior, Ablex Publishing, Norwood, NJ.

Mathematica in Education, Electronic Library of Science, Santa Clara, CA.

Mathematics and Computer Education, Old Bethpage, NY.

Mathematics Magazine, Mathematical Association of America, Washington, DC.

Mathematics Teacher (The), National Council of Teachers of Mathematics, Reston, VA.

Maths&Stats, CTI Centre for Mathematics and Statistics, Birmingham, UK.

MER Newsletter, Mathematicians and Education Reform Network, University of Illinois at Chicago, 1989–.

National Center on Postsecondary Teaching, Learning, & Assessment Newsletter, Pennsylvania State University, 1992–.

National Teaching & Learning Forum (The), (newsletter), Oryx Press, Phoenix, AZ, 1991–.

PRIMUS, U.S. Military Academy, West Point, NY.

Proceedings of the International Conference on Technology in Collegiate Mathematics, Addison-Wesley, 1988–.

Proceedings of the Society for the Psychology of Mathematics Education, Shell Centre, Nottingham, UK.

Project CALC Newsletter, D.C. Heath and Co., 1988–.

SIAM News, Society for Industrial and Applied Mathematics, Philadelphia, PA.

Teaching Professor (The), Center for the Study of Higher Education, Pennsylvania State University, 1987–.

UMAP Journal, Consortium for Mathematics and its Applications, Lexington, MA.

UME Trends, JPBM, Mathematical Association of America, 1989–.

Washington Center News, Washington Center for Improving the Quality of Undergraduate Education, Olympia, WA, 1984–.

Addresses

American Mathematical Association of Two-Year Colleges (AMATYC), State Technical Institute at Memphis, 5983 Macon Cove, Memphis, TN 38134.

American Mathematical Society (AMS), P.O. Box 6248, Providence, RI 02940–6248.

Association for the Advancement of Computing in Education, Box 2966, Charlottesville, VA 22902–2966.

Center for Teaching Effectiveness, University of Delaware, Newark, DE 19716.

Center for the Study of Higher Education, Pennsylvania State University, 403 South Allen St., Suite 104, University Park, PA 16801–5252.

Center for the Teaching and Learning of Mathematics, Box 3149, Framingham, MA 01701.

Consortium for Mathematics and its Applications (COMAP), 57 Bedford St., Suite 210, Lexington, MA 02173–4428.

CTI Centre for Mathematics and Statistics, The University of Birmingham, Birmingham B15 2TT, UK, e-mail: ctimath@bham.ac.uk.

Electronic Library of Science, 3600 Pruneridge Ave., Suite 200, Santa Clara, CA 95051.

Mathematical Association of America (MAA), 1529 Eighteenth Street NW, Washington, DC 20036–1385.

Mathematicians and Education Reform (MER) Network, University of Illinois at Chicago, Department of Mathematics, Statistics, and Computer Science (M/C 249), 851 S. Morgan St., Chicago, IL 60607–7045.

National Center for Research to Improve Postsecondary Teaching and Learning (NCRIPTAL), 2400 School of Education Building, The University of Michigan, Ann Arbor, MI 48109–1259.

National Center on Postsecondary Teaching, Learning, & Assessment, Pennsylvania State University, 403 South Allen St., Suite 104, University Park, PA 16801–5252.

National Council of Teachers of Mathematics (NCTM), 1906 Association Dr., Reston, VA 22901–1593.

Project Kaleidoscope, Suite 1205, 1730 Rhode Island Ave. NW, Washington, DC 20036.

Shell Centre for Mathematics Education, University of Nottingham, Nottingham NG7 2RD, UK.

Society for Industrial and Applied Mathematics (SIAM), 3600 University City Science Center, Philadelphia, PA 19104–2688.

U.S. Military Academy, Department of Mathematical Sciences, West Point, NY 10996.

Washington Center for Improving the Quality of Undergraduate Education, The Evergreen State College, Olympia, WA 98505.

Calculus on the Internet

Martin A. Flashman
HUMBOLT STATE UNIVERSITY

Internet resources related to calculus are growing rapidly. Access to many of these resources is available using three different techniques: FTP, Gopher, and the World Wide Web (WWW). These resources can be broken down into project descriptions, articles and other discursive materials about calculus, software (laboratory) materials, and course materials both commercial and non-commercial .

The simplest way to find many of these materials is through the Mathematics Archives, especially through its home page on the World Wide Web. The Mathematics Archives can be accessed by:

1. Anonymous FTP at `archives.math.utk.edu`.
2. Gopher to `archives.math.utk.edu` (with port 70).
3. WWW access using the URL `http://archives.math.utk.edu:80/`

An alternative to locate calculus resources through WWW is using search tools. MathSearch, which will search a collection of mathematical Web Locations, can be located at

`http://www.maths.usyd.edu.au:8000/MathSearch.html`

This can be particularly effective when things have changed as they will in what is available and where to find it. For instance, there is an interesting discussion of how the University of Tennessee is making changes in its calculus program found in the department newsletter at the WWW address of

`http://mathsun1.math.utk.edu/0h/Announcements/newsletter94.html`

If you want to join a discussion in a more interactive basis you might want to try a discussion list such as CALC-REFORM. To join this list send the following message to `listserv@e-math.ams.org`:

`SUBSCRIBE CALC-REFORM your name`

In response you will start to receive correspondence from around the United States and other parts of the world on many issues related to calculus reform. For a sampling of what this discussion has covered in a month or two, you can look at its archives which can be found using Gopher, looking at the Newsletters section at the Gopher site `gopher.maths.soton.ac.uk`.

These can also be found, along with many other mathematical Internet references, by using one of the general WWW homepages such as `http://e-math.ams.org/web`, the home page for the American Mathematical Society or `http://www.maa.org/`, the home page for the Mathematical Association of America.

What is certain is that access to information through the Internet will continue to develop so that this brief view will not be accurate in a few years or perhaps even months. However the changes occur, you can be sure that this technology will enhance both the faculty and student experience with calculus.